EXISTENCE

EXISTENCE

SCIENCE, SPIRITUALITY & THE SPACES BETWEEN

BRETT HAYWARD

Library and Archives Canada Cataloguing in Publication

Hayward, Brett, 1956–, author
Existence : science, spirituality, & the spaces between / Brett Hayward.

Includes bibliographical references and index.

ISBN 978-1-926991-11-5 (pbk.)

1. Life. 2. Meaning (Philosophy). 3. Science. 4. Spirituality. I. Title.

BD431.H39 2013 128 C2013-905513-4

Editor: Kyle Hawke
Illustrations: Tiffany Fok
Cover and Text Designer: Kyle Hawke
Indexer: Bookmark: Editing and Indexing
Photographs: Front Cover and Rear Cover photo 4 © 2014 Ade Barnett;
Rear Cover Photographs 1–3 & 5, public domain;
Vetruvian Man by Leonardo da Vinci, circa 1490, public domain

Granville Island Publishing Ltd.
212 – 1656 Duranleau St. Granville Island
Vancouver, BC, Canada V6H 3S4

604-688-0320 / 1-877-688-0320
info@granvilleislandpublishing.com
www.granvilleislandpublishing.com

First published in 2014
Printed in Canada on recycled paper

I dedicate this book to my children,
Anne Marie and Madelaine,
as a means of leaving a piece of myself behind.

Acknowledgements

The ideas for this book came out of my head, but without a good editor like Kyle Hawke it wouldn't have blossomed into this final product. Thank you, Kyle, for the ongoing course in correct English and open-minded diplomacy to effectively communicate to as many readers as possible.

There were other unnamed reader-editors, who were very insightful and maintained encouragement, while allowing me to fine-tune the language; their input was crucial. Thanks to artist Tiffany Fok for nicely standardizing my scratchy illustrations into a confluent whole. It might be a bit strange to acknowledge a society from five thousand years ago, but it was my learning of the Tuatha de Danaan, who possibly built the monolithic tomb at New Grange, Ireland, that clearly and precisely started this book, so they need to be credited.

Finally, thanks to my wife Deborah who allowed me the time to seclude myself in the man-cave for hours at a time, and occasionally pointed me back at the book when I wandered off to other distractions.

CONTENTS

1

Getting Started

YOU WHO READ THIS are a unique collection of one hundred trillion cells, all agreeing to work together, to intensely specialize into about two hundred different kinds of cells, conglomerating into organs, each with their specific job, so that you can exist.

Figure 1

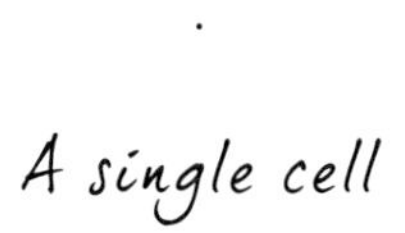

A single cell

You were originally one cell, just like a bacterium or an amoeba, when your mother and father each contributed half your DNA. Although just one cell, you were made up of billions of atoms that bonded together into complex structures like proteins, fats and nucleotides. These atoms were created long ago in a star, of which the Sun is the remnant, where intense heat allowed nuclear fusion to combine smaller atoms into bigger ones.

The final proportions of your atoms, like carbon, hydrogen, oxygen, sulphur and nitrogen, both now and when you were one cell, are approximately the same as those of the universe. As the saying goes, biology parallels geology — or to put it simply, life stepped out of the rocks. We take this for granted, mostly because we don't know how it happened, partly due to that kind of inquiry always spiralling into a fathomless vortex of mystery that leaves us unsatisfied. Add to that the complication of spirituality and the worldwide variety of viewpoints of God, and the big conversation about life shuts down.

When you were one cell, you started your life on Earth totally dependent on your mother for sustenance. The fertilized egg divided into two, then four, eight, sixteen, etc., until a ball of cells was produced that rolled around in the nutrient-rich elixir of the uterine fluids, gathering nutrients by simple absorption — much like the method that many tiny species of organisms still employ. When you became a little too big for the surface-absorption method of staying alive, a few millimetres in diameter, you adapted by sending down roots in the form of blood vessels, which eventually became the placenta, a clever interface you made between your mother's blood supply, with all its oxygen and nutrients, and your blood supply. Most mammals, from mice to elephants, do this but the interfaces differ.

As you developed from embryo to fetus, your body went through stages that many other vertebrates go through. The earlier the stage, the more difficult it was to tell the difference between you and a fish, a salamander, a frog, a chicken or a dog — an embryo looks like an embryo.

> I don't mind being compared to other species — it doesn't diminish my divinity, it augments theirs.

Near the end of your time in the womb, you grew hair. You weren't eating, so there were few feces, but your body had metabolism, so there were wastes from the kidneys, which emptied urine into the bladder. Your bladder connected to your navel, where the umbilicus took the urine to the placenta where your mother's blood took it away. The imagination runs wild with anecdotes about how long mothers have to deal with their children's waste.

Figure 2

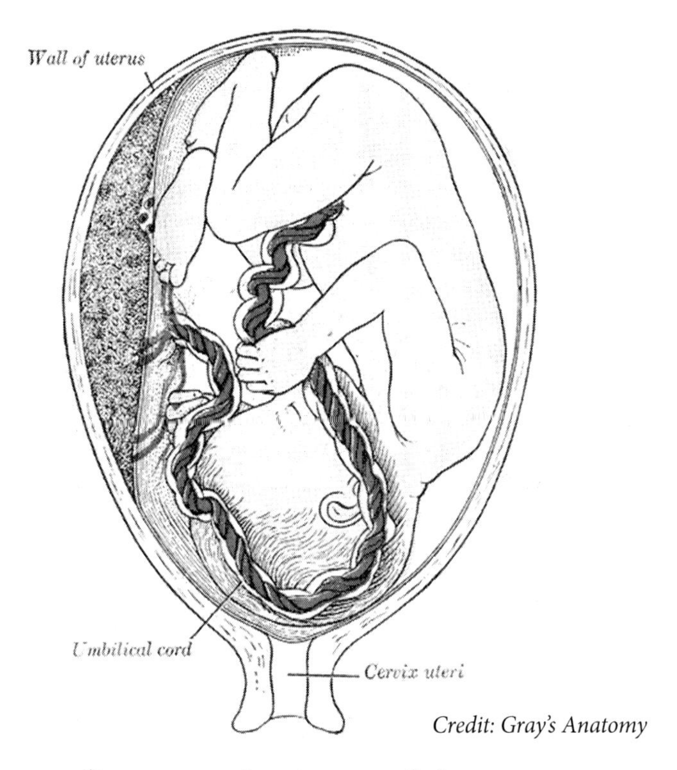

Credit: Gray's Anatomy

Diagram of a human fetus

So you grew and then you were born, and *that* was a freaky event. A grown woman, your mother, pushed out another complete living being into the world, an absolutely unique person, the likes of which has never been before and will never be again — even identical twins have different fingerprints.

Now, as you read this, you are located on a planet that is travelling 100,000 kilometres per hour in an elliptical orbit around the Sun, a smallish star about 1.4 million kilometres in diameter and 150 million kilometres away. We live on the

Earth's thin crust and breathe a thin atmosphere of oxygen. These celestial bodies, the Sun and the Earth, are on the edge of our galaxy, the Milky Way, which is spiralling around like a pinwheel, and it takes 250 million years to go all the way around. There are other galaxies that swirl around themselves as ours does and a whole bunch of them also rotate around some kind of centre point. Distances get mind-bogglingly big and are measured not in kilometres or miles but how far you would go in a year while travelling at the speed of light, 299,792 kilometres (or 186,000 miles) a second. Stars and planets are so big that the Earth becomes an unseen speck.

Figure 3

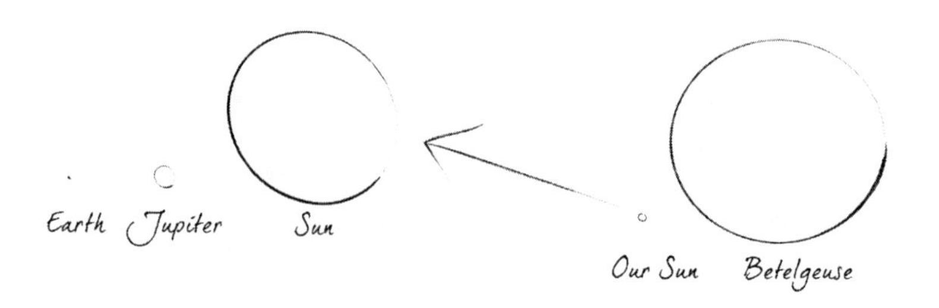

Size comparison of Earth to stars

So here we are — you and I, each 100 trillion cells — riding the thin crust on a speck of a planet in an incredibly enormous universe. We don't ponder it much because, like the musings of life itself, it makes us uncomfortable and gets us nowhere.

Some people ignore it all and just live and die, like the dinosaurs did for 150 million years. Some people look, through science, at how to unravel all the mysteries, to know and hopefully understand our place in it. Then there are some people who need to find meaning in life. I dip in and out of all three groups, but the last one is where I live: life has to have a purpose bigger than one that I imagine — something closer to a universal truth. This book is one viewpoint in that exploration. There is so much that I want to know, and so I've dared to ponder. The more I discovered, the more I realized how little I know, so there was often the exhilaration of discovery tempered by the humility of limitation.

This book honours the ancient Greek method of observation and contemplation as much as it uses repeatable evidence from science, but seeks to also include God. If we sincerely desire the truth of our universe, we have to have the courage to look everywhere, which would include spirituality. The ancient Greeks, from 600 to 300 BC, cracked our world wide open, but with the crumbling of the Roman Empire, Europeans mislaid most of the ancients' discoveries. Their knowledge was preserved in the Arab world and reintroduced to the Europeans through the expansion of the Moorish empire into Spain and by the invitation to Arabs by Italian scholars. After more than a thousand years of somnolence and ignorance, the re-awakening of the exploratory mind began with translations of Greek knowledge from Arabic to Latin and Irish Gaelic.

Extinction or Choice?

Scientists believe that 98 percent of the species that have existed on Earth have gone extinct. The process by which extinction is avoided, if only temporarily, is adaptation. This term does not denote that the winner is the fastest, or smartest, or strongest, or richest, but simply the group that had the ability to adjust to the changing circumstances to survive and thrive. Evolution only describes the genetic changes that occur and the selection of the most adaptive beings, but many people have expanded it to include the origin of life — which, as we shall see later, is not correct and confuses the issue.

It is in our best interests to look at the past to see the trends and assess what has worked in order to project where our best path into the future lies. The tricky part is that we don't know what the next catastrophe will be that tests our viability as a species.

Is humankind at the mercy of evolution or can we make choices that are apart from that process? It might very well be that, just as one generation has to die to make way for the next one, our species *must* move aside to make way for the next hominid. There isn't much genetic variation between us people for a stressor to test our viability, so we are all in the same species boat and our social adaptability could be the single greatest variable to survival. Our current way of being has followed evolution in that all of the beauty and horror of humankind has arisen in a free-for-all way, with 'survival of the fittest' governing who comes out on top, from researchers to violinists to politicians to athletes to painters to generals, you name it.

The overseer of this whole process has often been *force* in some way, and the net result of all those individual forces is the world of us *Homo sapiens*, the most complex and world-changing species we know of. Still, our numbers and activities have expanded without limit and are headed for collapse so, if we as a species plan on surviving on this planet, we need to invent new ways of being. Denial runs heavy, as most of us believe that the great *Homo sapiens* will find new answers to huge problems and so we don't need to do anything. This was the path that Neanderthals took and the last living member of that group, on the northern shore of the Mediterranean Sea about 25,000 years ago, might have had some advice for us.

What Is Our Species?

Homo sapiens by name, but with lots of other descriptors:

Human race

There are races of people amongst humans, and humans are a species amongst other species, but to call the species a race is confusing. Maybe it's a pun on our busyness.

Human

It denotes us as hominids that are more than animals, *Homo sapiens* with thinking and feeling.

Human beings

This looks like another layer of appreciation on top of 'human', which adds pomp and ceremony, like when spoken from the pulpit or lectern. Are there horse beings and fish beings?

Man

Over the past 200,000 years there have been some 85 billion *Homo sapiens* who have walked the Earth, and half of them were female, but women only got to vote in the last 100 years, so it's obvious how the whole species has been denoted as 'Man'. The species of horse isn't called Stallion, nor pigs Boar, nor chickens Cock. Wait a minute. No, forget it. 'Man' is a species term that can be retired.

Mankind

Based on the evidence of the horrors we've done to each other, to other living creatures and to the planet, this is clearly an oxymoron that contains more hope than fact.

People

This is a warmer term that elevates the species into personable individuals, rather like referring to us as men and women, instead of the more categorically correct male and female, which appears on hospital washroom doors and on insect identifier pins.

Humankind

This is a more workable expression that denotes perhaps if we're more human we'll be kinder, and that our kind of hominid is human.

So, *Homo sapiens* is the species, made more personable by the term 'human', and when grouped together referred to as 'people', and to include every kind of people, we are 'humankind'. Tidy. As *Homo neanderthalis* earned the nickname 'Neanderthal', we can get personal with our species by adopting the nickname 'Sapiens' — which means 'wise', by the way.

The Storyline

In the process of inquiring into how life works in order to prophesy which road we should take in the near future (or if we have no choice, just to understand how we came to be on the road to our destiny), it was necessary to roll time backwards, unravelling sequences back to their beginnings to understand how things built up. Back to the Big Bang and the very beginning of the material world, back to the Little Bang, DNA, the start of life and back to the Big Event, where Sapiens attained consciousness.

Figure 4

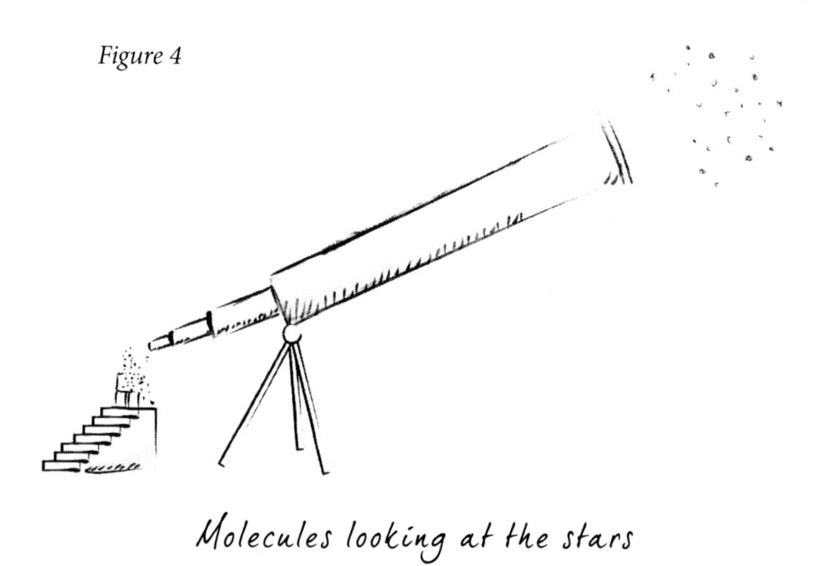

Molecules looking at the stars

While there have been many crucial turning points, both before and after this Big Event, it was consciousness coming into being that defines who we are as a species — *the becoming aware of time and space, of mortality and God.* It is the hallmark of our leaving the animal world. It must have happened to one person in an instant and spread from that moment. From afar in time, the event could look succinct, but the closer we look, the more likely there were really several smaller events. The elegant poetry of a single event would not be lost entirely by its splitting into smaller divisions, because the process would still start with a single moment in an individual person who became aware, for the first time on Earth, of existence. For the first time ever, since the rocks washed into a slurry that produced life, life started looking back and wondering how and why.

This inquiry into existence includes spirituality and God, while touching upon some of the scientific disciplines such as biology, astronomy, chemistry and physics. Many people see spirituality and science as totally opposite but — in actuality — it's sometimes difficult to separate them, because, as we will see later with matter and energy, one without the other may be meaningless.

2

Molecules and Tiny Stuff

What a Ride!

My wife Deborah and I were snuggled in bed one dark winter night when it was raining lightly, and we could hear the dripping of water from the roof into the downpipe. I imagined riding the drop that had fallen from the sky to be collected here, dripping down into the earth, to percolate into the porous spaces of the soil and to run in progressively larger rivulets to the sea. There, the drop could partake in a million different ways of being — to be swallowed by a fish or become part of seaweed and eventually be passed on into the sky to become part of a cloud again. What a cool ride!

Now, I reflect that I am made of countless molecules of water that have come from everywhere, that ride in me for a while and then are passed on. For a moment I realize that I am part of the cool ride for, although I am living, I am part of the molecular exchange that has been happening on Earth for billions of years. It is amazing that this cluster of recycled atoms can write to you about this viewpoint. Galaxies, stars and planets don't write. Hydrogen, oxygen and carbon don't write. But here we are, in the middle, writing and reading.

Along with this awareness is our consciousness about existence which we see, as far as we know, that no other collection of molecules has. Our ponderings about life are unique and so far we haven't found anyone else in the universe that shares this attribute with us.

Creation of Life

Molecules were created in our star, cooled to form the Earth, then washed by water into a solution to form the perfect circumstances for life to arise. There has to be a driving force to make molecules stay together, reorganize themselves consistently to make life occur and continue when everything inanimate is moving apart to obey entropy, or maximum chaos, when galaxies are spinning farther from each other and any neutral atom released into liquid or gas will maximize its distance from other atoms. In fact, if it wasn't for gravity, everything might have blown apart from the very start.

Gravity, as we commonly know it, is the force that brings us down when we jump up, and affects every molecule in tiny amounts, which accumulates into massive forces when conglomerated together in planets and stars. We don't know how it arises or where exactly it comes from. It is always positive and has no opposite force, which is very odd as far as our understanding of the universe goes.

The creation and existence of life takes energy and organization — one without the other precludes life. Our solar system (and the whole universe) is winding down in energy, but we are talking billions of years, so not to worry. The 'closed' system of the Earth and the Sun — as a package — is decreasing in energy, and the Earth, as its own closed system, has a finite number of molecules. When viewed objectively, this cooling Earth-Sun package managed to muster up life, which is an uphill endeavour — it takes energy and organization. The Second Law of Thermodynamics says, "In any closed system, the entropy of the system will either remain constant or increase," with entropy being chaos or maximized freedom of movement. So though we might expect less organization and more chaos, life arises. What was the driving force for life to come into being and why does it exist? Science has no answer. The best that science can give so far is an open-minded lack of conclusion because its answer hangs on uncertainty and randomness. While it is true that all the factors necessary for life were there waiting (namely chemicals, water and the right temperature/pressure/energy levels), for them to magically leap from a vat of warm mush into life required a missing piece that we haven't even touched upon yet. All the ingredients are there and the force to mix them — but without a guiding force to create something out of chaos.

Repeated Ideas

I heard someone quip a few decades ago that the likelihood of life arising spontaneously from all the right conditions of water, molecules, heat, pH and randomness was as likely as a tornado going through a trailer park and producing a jumbo jet. The analogy presents the scenario of building blocks, force and chance as the beginnings of life and does not mention God or some kind of intelligence, although religious people have used it in their argument for the existence of God.

There are many permutations about the jumbo jet analogy, by many different authors, but they all lead back to Fred Hoyle giving a radio lecture in 1982. What he said was, "A junkyard contains all the bits and pieces of a Boeing 747, dismembered and in disarray. A whirlwind happens to blow through the yard. What is the chance that after its passage a fully assembled 747, ready to fly, will be found standing there?" He was speaking statistically.

This is reminiscent of the Watchmaker Analogy that William Paley wrote in a book in 1802, that describes someone finding a watch in a field and hypothesizing that its complexity conclusively means that someone made it. The trail doesn't end there — there are pieces of history that suggest that scholars were discussing the same concept a hundred and more years earlier. Leonardus Lessius (1554–1623), a Flemish Jesuit, used the analogy that the intricate details of nature couldn't have occurred randomly any more than a beautiful palace could come into being on its own or by chance.

This can raise questions, not so much about God or evolution and all those arguments, but about whether Hoyle had heard of Paley's ideas, or whether Paley had heard of Lessius's. For when an idea arises spontaneously and repeatedly, you have to wonder about the source of those ideas.

Ancient Contemplations

Our ancient ancestors, back to the dawn of civilization, had the same genetics as us and they had consciousness, and it's not hard to imagine that amongst them were individuals who contemplated the universe, life and destiny. One of the earliest written records of this is from the first few pages of the Bible, probably written about 3,500 years ago, when the writer says that "the Lord God formed man of dust from the ground, and breathed into his nostrils the breath of life; and man became a living being." For the simple folk who lived back then, this was a clear picture of how life came into being, satisfying their yearning for purpose and understanding of life. If you look at it objectively and remove God from the picture for just a moment, the story is very similar to the scientific one, on the previous page, with viewpoints from physics and biochemistry: chemicals in the earth became life.

I know it's a bit weird to put a capital letter in the middle of a word, but I use 'nonGod' to differentiate the generic term 'god' from the monotheistic belief in a God — 'unbeliever' or 'non-believer' seems like 'less than'. Everyone believes in *something* that reflects their humanness.

The biblical story places God as the source of the force that bridges the gap. For the nonGod group, ascribing cause and effect to an unseen, unproven God is not reasonable, and those who don't hold this viewpoint can still respect it. They prefer to believe that science will one day tell us what that force is. There are those who, at this time, prefer not to wonder and not to know. The God group should not worry about that or see it as a challenge to their faith. In our inquiry, it's fundamentally important for us to accept others' beliefs as an essential part of the pool of possibilities. Any one viewpoint cannot see all the paradigms, and insisting that ours is the correct one is unnecessary. The truth will eventually speak for itself.

Hidden Messages in Water

Many of you have probably heard of the discovery by a Japanese fellow named Masaru Emoto, who observed that by exposing pure, distilled water to different words, emotions or music and then freezing the water, different crystals resulted. Beautiful words and music created beautiful, organized crystals, while anger, swear words and discordant emotions created stunted, coarse crystals. There are many diverse ideas that have arisen from this,[1] but the point here is to grasp the concept that inanimate things like water molecules can reverberate in response to how they are treated. If that is true, then the water in our bodies could be susceptible to the same phenomenon, whereby negative emotions could create negative molecular energy. If water can respond this way, perhaps all molecules can.

1. Including that other researchers have not been able to replicate results; Emoto has noted that his selection process "is not strictly in accordance with the scientific method."

From here, your mind might expand with possibilities about the relationship between molecules and life, where life is our point of view and we are made of molecules which might have a life of their own. When we care for our bodies and each other, we could be caring for immortal pieces of the universe because we are made of eternal molecules. Also, water gives an unbiased opinion as to what good and bad are. We could argue that those precepts are merely points of philosophy or social mores accumulated by a series of successful observations or decisions, but when molecules come along and show us their response to our behaviours, we have an impartial arbiter.

"Water is angry at us," Emoto says. In an interview with Wendy Schuman of Beliefnet, he said the message from water is, "Be aware of giving love and thanks. If you can do that, the water will be happy and it will shine within. And that will help bring happiness into your own life. Don't be selfish, because if human beings are selfish, water will be selfish. And that will create an imbalance of energy. That imbalance will pile up and finally burst into destruction. But I understand that it is difficult to have that idea of love and thanks all the time. So just being conscious about it makes a difference."

This simple philosophy overlaps with religions whereby loving, gratitude and generosity are some of the major factors attributed to bringing happiness into our lives. Emoto doesn't get into where love comes from, why to be thankful or how to be generous — the conversation of spirituality — and so we could take his viewpoint as a non-religious guideline for better living based on what water has taught him. If so, then water has desires similar to what we understand those of God to be, and you don't have to believe in either of their viewpoints to see that they are true.

What Does It Matter?

A high school physics teacher once remarked that matter is so porous that we should be able to walk through rock. He was referring to how small a nucleus is relative to the size of an atom. A black hole or neutron star can collapse matter into a compact state, so that an enormous body like a star can collapse to the size of a suitcase.

Here is a molecule of ethyl alcohol, CH_3-CHOH, showing the nuclei of the atoms without the electrons:

Figure 5

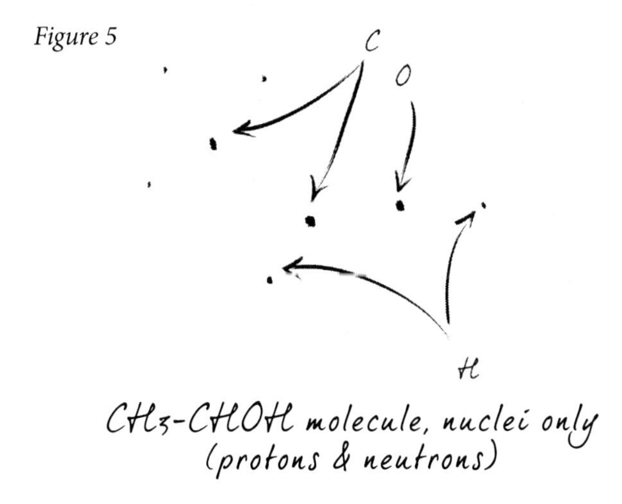

CH3-CHOH molecule, nuclei only
(protons & neutrons)

Here it is with electrons whizzing randomly around orbits so fast that there is a sphere of electrical energy around them:

Figure 6

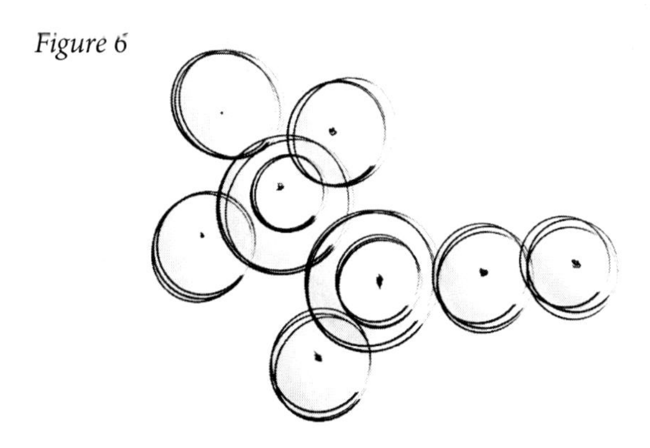

CH3-CHOH molecule, electrons added in

This is the functional unit with energy giving the shape and volume, and the tiny amount of mass at the core giving it its weight. So this is how it 'looks' when it is all poofed up and ready for action:

Figure 7

CH_3-CHOH molecule, 3D physical representation

And this is how it looks in a black hole:

Figure 8

CH_3-CHOH molecule, collapsed in a black hole

When you consider the relativity equation, $E=mc^2$, E is energy, m is mass and c is the constant, which is the speed of light, which is a really big number and so m has to be really small to make the math work. This means that a tiny amount of mass can become a huge amount of energy, an example of which is the nuclear bomb. It also delineates that mass is only a small part of the universe, as illustrated above, with the dots of mass made functionally huge by the energetic, massless electrons whirling around their opposite charge in the nucleus.

The energy component is so huge and the mass part so small that it would seem that mass is simply an anchor for energy to find a focal point to exist. Without that anchor, energy would be totally erratic and chaotic. Mass without energy is absolutely useless as far as life is concerned — a black hole doesn't participate as a star or a planet but as a gigantic, overendowed gravitational toilet. So although we look at material things as solid, and our reality wraps around that idea, most of our world is energy. Mass and energy need each other and neither one alone has any meaning. This is a fundamental pillar of the idea that unity and cooperation can create more through synergy than either factor can alone, with implications for human relations, our species' survival and life itself.

Universal Things, Light Cooperates Too

Light has been described as both a wave form and a particle, depending on what aspect of it you are focused on. Whenever there is a mix of explanations for an event or thing in scientific disciplines, it usually signifies that we haven't yet truly discovered what it is all about. So it is with light — we don't yet fully understand it. We know lots about it and can predict many of its behaviours, but have not seen it clearly. As observers, we have to confess that we might not have the observational equipment to view it in its fullness or that our very finite brains cannot absorb what it is.

Sometimes there are phenomena in nature which are repeated at different scales of size, and they help us to understand the universe, and so we could look to waves in order to seek insights into light. The waves of the oceans have come to shore for billions of years. On a sandy beach, you can see their daily mark when the tide goes out.

We intuitively grasp at a glance that the water has pushed against the sand, the sand presented some resistance, and that momentary pressure of water against sand created a small ridge. The tide kept coming in, the water got deeper then flowed over the ridge, only to be met by another slightly rising plateau of sand ready to offer resistance again. These are two media that somewhat neutralize each other for a moment until one breaks out of the resistance.

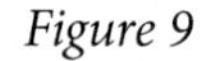

Figure 9

Sand waves

Another common sight, which could be on the same day that you are looking at the sand at your feet, is up in the sky, in the clouds, which can also form waves.

The two media here are bodies of air, where the lower, warmer air is flowing along, rising, and meets the cooler, more stable air in the upper atmosphere. As the rising air cools, the water vapour condenses to form clouds, and the upper air won't budge. After a moment of resistance, the lower air shears down, only to rise again and condense into cloud to illustrate the meeting of the two bodies of air. The air skips along, leaving the legacy of the clouds.

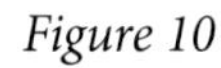

Clouds

In both these cases, we see that the wave is not the mark that is left but the force or action that creates the mark. We also see that it takes two opposing media to create the mark of a wave.

That might be how these marks appear to have formed, but a third example gives us some more information. When watching a quiet lake surface, we can see the waves created by a passing breeze. As with the first two examples, there are two media meeting, in this case the air meeting the water. However, this time we can see the formation of the waves, not just marks in the sand or the clouds. All the little waves have the same height and space between them, which you wouldn't expect if they were caused solely by the wind, as the wind would likely hit the water with variable speeds, causing various sizes of waves. It appears that the water has an inherent ability to resonate, and once touched by the energy of the wind, it does so on its own. The wind was needed to start the wave action but once that force was applied, the water responded in a set, geometric way, like a tuning fork. Thus the sand on the beach simply recorded the wave action that was in the water, and the clouds formed from the air rising and condensing into the peaks of the upper air's wave form.

So the resultant mark of light might be the meeting of the *energy* of light with 'something' else. After reading about the concepts that physicists are now showing us, we can begin to think of space as no longer emptiness but the elixir of the universe, of 'something' — dark energy, or particles coming into and going out of existence — rather than 'nothing'. So it could be that space is this uniform milieu that light is bumping against to make a wave form. Similarly, household electricity is basically electrons going along a wire one way, generating a current the other way, so no thing, no matter really moves but an effect is seen at the end of the wire. Perhaps, then, light is not a thing that goes from here to there but an energy form that passes itself from particle or antiparticle to particle, 'phosphorescing' them temporarily, so that it looks like wave and acts like particle, when really it isn't any 'thing' but the result of that energy form. As we have seen with the balance within atoms, where mass and energy commune, and with electricity, where energy and matter are organized together, it must be that light is energy but it is insignificant without partnering with matter. This could add credibility to the idea that light waves exist as a result of the interplay between the energy of light with some unknown matter. You could speculate that light could go at limitless speeds if it didn't have this unknown matter slowing it down, and on the other hand, maybe without these particles, light would go nowhere.[2] If any of this has merit, it could help explain other ideas we have that are confusing, like the concept

2. I have this hunch that the *speed of light* is telling us something of this cooperative 'unknown' that light uses to get around.

that light can bend. If we observe light bending around an eclipse phenomenon, perhaps it is these particles of space that are subject to gravity that are being pulled, and the light, just following the path of least resistance, follows the particles that have been pulled off the path. Then again, if spacetime is curved, then light, which is part of that parcel, should certainly be able to curve as well. Some experiments with light pass it through a vacuum to prove theories. A valid question would be, "Is that vacuum in the lab the same as the one in outer space?" Which is to say, is our manmade vacuum full of stuff too?

The other idea that goes with this is that light is just a portion of the electromagnetic spectrum that comes from a source, and is mostly relevant to those who have eyes and can see or to plants that can absorb the energy and convert it to mass. There are also gamma rays, X-rays and cosmic rays, and the difference between them all is their wavelength, which would be a reflection of their energy levels. So the more energetic forms of light with higher frequencies would be less deflected by the particles, like a fighter jet cutting through the storm compared to a single engine airplane getting bounced around.

So they're really a family of waves and, if we find out what a wave is, it will answer questions across the whole family. If we can ever elucidate a Grand Unification Theory, GUT, that bridges Newton's physics of celestial bodies with Einstein's physics of atoms, it will include light, like current models, as well as the orphan force of gravity that hasn't yet fit in anywhere.

Size Matters

At the end of the movie *Men in Black*, we see our universe sequestered into a clear ball that some strange aliens are playing a game with. Similarly, at the end of the movie *How the Grinch Stole Christmas*, we see that the entire venue of the story took place on a snowflake. So the concept that there are worlds within worlds is a common one, if not a logical one. Michio Kaku explains quite well in his book *Physics of the Impossible* that things change drastically as size changes and that the laws of physics change: ". . . scale law rules out the familiar idea of worlds-within-worlds." So our imaginations can stop at the idea of a universe in a drop of water, or our universe being a tiny part of something else.

The First Particle

The most helpful and profound theories in physics have been those that are simple and elegant. When looking at the basic building blocks of the universe and life, we started with the atom as the tiniest, most basic thing. Then the atom turned out to be made of neutrons, protons and electrons. Now, our minds are being pushed to understand that protons and neutrons are composed of still smaller things called quarks. There are three different 'colours' (they're not coloured at all, it's just a descriptor) and six different *kinds* of quarks. At this point, it looks like you need a Field Guide to Modern Quarks to navigate *inside* an atom. Elegance has been lost but not the pursuit of the secrets of the universe, because the exercise might prove to be the preparation for another paradigm leap. As the quark story gets progressively more complicated, it is probably the cerebral stretching in preparation for the elegance to come. Other things in physics *appear* different when viewed at different angles, times or states. Water can be ice, liquid or vapour, and so quarks might prove to be simpler than we thought.

At any rate, if the quark is the simplest building block of the universe, as Stephen Hawking believes, then we are at the end of the spool of matter and can get on with finding out what brought the quark into being. If eternity truly exists, then within that framework quarks could have always existed. Absolutely everything else in nature and the universe seems to be always changing, always in flux, morphing into something else, so why would quarks be any different? Also, our process of thought has systematically revealed how the universe works, by dissecting down to smaller and smaller building blocks, and each step of the way a 'block' had to exist before it could combine to create the larger 'block'. *If* that process continues throughout creation, back to the beginning of matter and energy, then it must be that quarks didn't exist and then they did.

Sometimes there comes an event or phenomenon that is a distinct change from what came before, or a turning point after which things were never as before, due to a tipping point, passing a threshold, critical mass or singularity. Regardless of whether we know the cause or not, these changes are labelled gateways to signify a door opening or a portal at a discrete point in time. Gateways would include consciousness, the wheel, metallurgy, sailing, writing, light optics in telescopes and microscopes, and electronics.

Our imagination deduces that at some point there is a gateway beyond which no tangible particle existed. So perhaps quarks have always existed but, for this conversation, we will wrestle with the possibility that at one point they did not.

We arrive at this first quark by reversing the forces of the universe, sending stars and planets and galaxies towards each other, back to a melting sphere of matter and energy. From there, we go back to incredibly high temperatures where nuclear fusion and the creation and annihilation of innumerable particles reverse down into just hydrogen, the simplest of all atoms, and ultimately into the building blocks of atoms, the quarks, further and further back in time. This is the critical turning point in our understanding of the universe — what happened for that first quark to exist?

After the first came the second, then zillions more in a billionth of a second, in popcorn-machine repetitiveness that churned out quarks in the form of hydrogen (H) nuclei (made of three quarks) and, a billionth of a second later, helium (He) atoms made of hydrogen nuclei fused together by high heat. There you have one view of the Big Bang — zillions of tons of H and He exploding apart, creating heat and starting the expanding universe, where all elements are created, where gravity causes random collapse of atoms into planets, stars and galaxies.

It may have started with one quark.

We could theorize that energy, pure energy, made the leap into matter: $E=mc^2$ or $m=E/c^2$ and that window opened for a moment, creating the zillions of quarks in the universe, then closed.

Why did that window open? Why did it close? Or rather, what made the window open and close?

When we find the perfect mathematical equation that describes the opening and closing of that window, will we have arrived? Will that answer the longing of our existence and give us rest for our souls?

Electrons

Physicists tell us that all electrons have the same amount of mass or rest energy. Observation of everything else in nature might suggest to us that there is variation amongst them. Maybe the variability is small, so small that we don't have the ability to measure the differences.

Figure 11

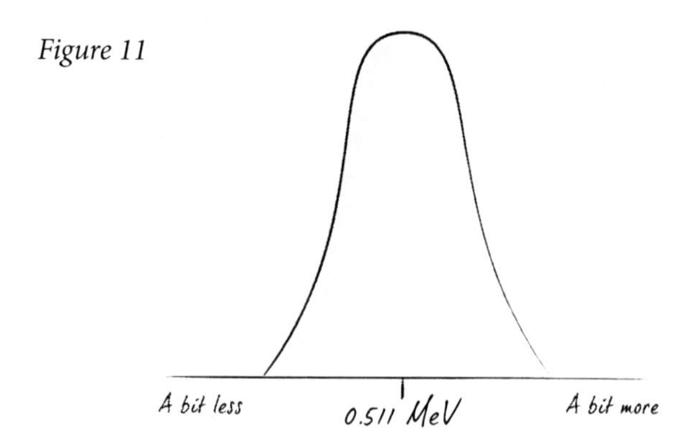

A bell curve diagrammatically describes phenomena, like observations of a population or quantity at a particular point in time – in this case, the energy of electrons in 2012

If there *is* variation, then there is the possibility of random aggregation of similarly charged electrons, either a bit more energetic or a bit less energetic, which then makes atoms or molecules they are a part of become dissimilar to others of the same construction. Therein lies another asymmetry to nature and the universe, which would explain how there can be physics 'constants' yet still variability in universal phenomena.

In the realm of electrons all being identical, there is the theory that there is only one electron in the whole universe. It is a concept that is difficult to understand, as the first obstacle for the mind to overcome is the incredible speeds necessary for an electron to be everywhere at once. It reminds us of God, who is everywhere at once, unfettered by time or space, and gives us a glimpse of the possibility of material and spiritual worlds overlapping.

They're All the Same!

There is an evolutionary tree for life but you have to wonder whether there is one for energy as well. Nature is very expedient and conservative, so it is likely that in the beginning there was one kind of particle. As fantastically high temperatures existed in the first few seconds of the existence of matter, the initial building blocks could have fused in different ways, creating new entities. Some of these new particles could have existed for a time to serve a purpose, then died out (reverted to energy or were annihilated) or became part of another particle — or something we haven't thought of yet. So perhaps the GUT has to be applied to the initial particles. Then again, that could be meaningless, because we already know that at very high temperatures all particles attain the same energy — but does that mean that they become the same or do they maintain their distinctness?

Life and its original creation is an extension of the unknown driving force of Nature to be 'more'. It is more primeval than DNA, because the original molecules came together before cells were formed. It bespeaks the possibilities of molecular intelligence or an external guiding force. Perhaps that force granted that intelligence. No matter how fine you slice or tune the questions, it always comes down to the possibility of innate things either acting on their own or being guided by an external force. Many people see God as that external force and so, even as science zooms in to smaller and smaller options, it still refines to the same argument but in a tinier forum. The calculus curve comes to mind:

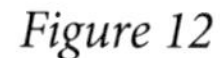

Figure 12

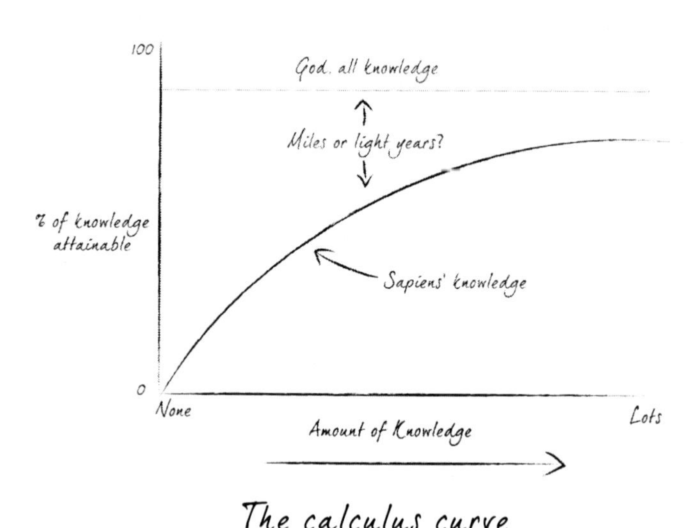

In this graph, the top line is all the knowledge that exists. From the God group's perspective, this line is God. The graph of the rising curve that can be described by calculus is the accumulated knowledge that humankind is attaining. The mathematical description of this curve is that it always comes closer to the top line but never quite gets there. When this concept is discussed with scientists who are aware, they will hold their hands apart to illustrate the width of the gap between what we know and the total knowledge that is out there. It seems that the more the scientist knows, the farther apart he stretches his hands. We delve into the fringe of knowledge and the amount of what we don't know that we don't know could be exponentially greater. The purpose of viewing ourselves in that light is to uncover the exponential knowledge, not to denigrate ourselves.

At each diminishing arena, the observer can elect to choose whether science is taking us to where we want to go and whether to entertain the possibilities of a relationship with God. Science offers the tantalizing promissory future of a still finer answer. Experiences in the realm of spirituality suggest that it can hold some sort of completion for us, an understanding of where we fit into the universe. It would be nice to pursue each path to its endpoint, and you have to wonder whether the paths are parallel, unrelated or converging to the same conclusion.

One coach in the quandary was Jesus, who gave us the clue "That to enter the kingdom of heaven you must become as innocent as these little children." Wrapped up in that idea is the notion of trust, which is the basis of intimacy in relationships, and so Jesus helps us set our minds to a path that might get us intimate with God. At first glance, this type of innocence doesn't appear to have much use on the science path but perhaps there is an opening of the mind, regardless of current logic, and a freedom from criticism and condemnation where a cerebral innocence could aid in scientific breakthroughs.

There are many theories, faiths and viewpoints around spirituality, but most of them contain the essence of a material world that binds us with limitations alongside the idea that when we are no longer matter but are pure spirit that we will be with God, or in some way released from limits. If this is true, then the calculus curve (Figure 12) applies to our path in science but not in spirituality — the bottom curve comes increasingly closer to the goal of the top line, but never quite gets there, with 'closer' being a relative term that could mean very close or light years apart. Even if science cannot reach God, it might point the way, if only by exhausting all the avenues that did not point the way.

If we think of our complete 'world' as a melding of the material and the spiritual, worshipping God helps to unravel the mysteries of the spiritual world but doesn't directly help us solve the mysteries of the physical universe. On the other hand, our scientific advances make our daily lives more comfortable and secure, which seems to correlate with diminishing faith, at least in the developed world.

Can we not choose both, and have external comfort from science and internal comfort from faith?

3

Cells and Biology

Life Continues

Individual living cells are the building blocks of life, the first level of intelligent organization of molecules into reproducible forms and the initial bridge between inanimate chemicals and life itself.

When you look at a food chain of life, you see the familiar sequence of command — on top are the carnivores, who eat the herbivores, who eat the plants, which come from the soil, where bacteria live.

Humankind is above all this, the grand commander of life on Earth, dominating and controlling everything beneath us. There are other species who are at the top of their chain as well, and Sapiens holds but one place in the enormous and complex web of interlocking chains.

Another way of looking at the organization of life is the taxonomic or species tree of life. There are three major classifications of living things, or domains as they're called, and you might think that they would be animals, plants and then smaller things, like insects and bacteria, but no.

The three types of living things are based on the types of cells and the genetics within them. These classifications are Archaea, Bacteria and Eukaryota.

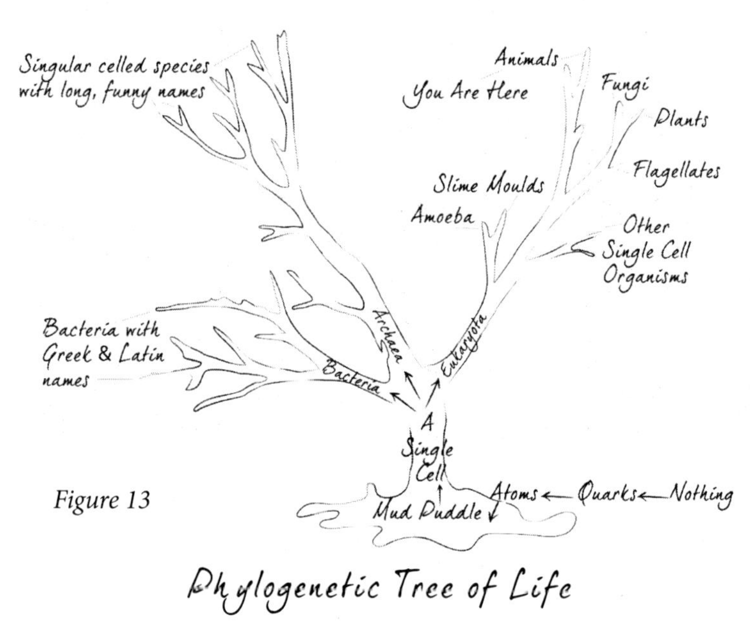

Figure 13

ARCHAEA are intriguing in that they are tiny, single-celled beings that are incredibly simple and yet are amazingly adaptive to horrible environments. It could be that their simple design makes them more hardy for the harsh situations that they can thrive in — the closer a being is to the non-life state of molecules, the better able it is to survive like a molecule. Conversely, the more specialized for a particular niche of temperature, moisture, nutrients and gases that a species is, the more fragile it is amongst the variations in environments on Earth. For examples, neither a tiger nor kelp can live in the desert, in Antarctica or by volcanic magma.

Archaea can be found by undersea heat vents, where high temperatures result from leakage of gasses from deep in the Earth, reminiscent of our understanding of the extreme conditions that existed when the Earth was first formed. They lack the more sophisticated internal machinery that animal cells have and they are capable of using alternate sources of energy besides sunlight, so some biologists believe that they are the first inhabitants on Earth.

BACTERIA are also single-celled creatures that are found in all kinds of places. They have different metabolic ways of being and different genetic sequences that make them stand apart from either Archaea or Eukaryota. They show amazing diversity and, amongst the thousands of species that exist, they can be free living, or a symbiotic element essential for other lifeforms to exist, or nasty pathogens that cause disease. Many can form colonies that then specialize to different shapes and functions, almost as though practicing to be a multicellular organism. Some biologists believe that this domain came first, then the other two.

Finally there are EUKARYOTA or the 'true' celled beings. If a being is not a primitive, tough Archaea or a ubiquitous Bacterium, then it is in the Eukaryota domain. This would be absolutely everything else living: trees, grasses, slime moulds, elephants, ants, all the fish in the seas, you and me, dogs, algae, turtles, amoebas, bees, all the birds of the world — you get the picture. For most of us, we easily recognize the animals around us as living creatures — the cat, the dog, the budgie, the fish in the tank or a bird flying by outside. We might consider that the house plants and the trees outside are alive, but they mostly form part of the landscape without drawing our attention. So, in addition to all the people, this is what we consider life on Earth. It's our daily paradigm.

Still, when you consider the total current mass of the three domains described above — known as the biomass — and the history of the Earth, you see a bigger picture. Looking at all the forests and jungles, all the oceans, rivers and lakes, all the mountains and deserts, and all the living species that live in any of those niches, you get the total biomass, and the Archaea and Bacteria domains comprise about half of it. One half of the weight of all life on Earth is microbial. There is more mass of krill (which looks like a little shrimp) than people. There are probably more than ten times the mass of ants than people. The oceans' phytoplankton are single-celled algae that produce three-quarters of the oxygen in our atmosphere. This is all to say that much of life on Earth is invisible to us, microbial, but comprising more individuals and mass than all of the other living beings in our world, including people.

There is a big contrast between our viewpoint that this world is *ours* with a few dogs, cats and birds thrown in, along with some trees and bushes, to one where we are the tiniest minority. Apparently, for three billion years the microbes of Archaea and Bacteria were alone on Earth with nothing bigger, such as a simple multicellular organism, to proclaim themselves sovereign over everyone else, as we do. This is the way of the Earth and Mother Nature — and all the universe — where there are always reserve pools, in this case lifeforms, where the microbes contain such hardiness and diversity to be the safety net for the existence of life on Earth. Invariably, there is a vast resource from which springs the latest evolutionary advance, both seemingly necessary to attain some critical mass and also to be the pool from which other creations could emanate, should the previous trials fail. It makes you wonder about the destiny of our species and possible future hominids.

From Chemicals to Cells

Viruses

Viruses were not mentioned in the description of life because many scientists do not consider them 'alive'. The current definition of life includes things such as growth and death, being able to reproduce, responding to stimuli, etc. Inanimate things like rock or water can do some of these things but to be considered alive you have to be able to do them *all*. Viruses don't respond like other living creatures do, which is one key argument that keeps them out of the classification of life, but there are many types of viruses that have evolved to infect all kinds of tissues and all kinds of species, including animals, plants and even bacteria. Viruses don't respond as individuals but they respond *as a group* or species by adapting to many kinds of tissues and species. That is to say, they don't respond to light like a worm or a sunflower plant, but move into select cells en masse, which is a response. They inject their DNA or RNA into a host cell and the cell uses its own machinery to run off more viral DNA — so a virus cannot 'live' without a host, which is another argument against them being truly alive. We humans can't live without all the bacteria in our bodies. Ticks, fleas and mosquitoes can't live without a blood meal — for most of them, if there was suddenly no blood on Earth they would go extinct. So there are different definitions of interdependency. Viruses are closer to being a set series of chemical reactions, bonds and assemblages than they are to being creatures and, as such, they are a fine bridge between the inanimate chemistry of rocks and all animate living things. Part of what they are illustrates what Nature was trying to do with chemicals before life arose. A notion that supports this is that some viruses have their genetic material as RNA, whereas all other lifeforms have it as DNA, suggesting that RNA might have been the earlier step to preserving reproduction information. Viruses show that molecules can consistently congregate in a set way without meeting the strict definition of being 'alive'. It again brings to the fore the idea that there is a force that is pushing chemicals to aggregate, to become complex instead of chaotic, to reproduce themselves in a consistent, orderly fashion whereas, outside of the world of the living, chemicals have no such need.

DNA

One day long ago on this little planet of ours, DNA — or the 'Little Bang' — came into being. All other molecules in the universe either prefer to distance themselves whenever possible (maximum entropy) or they are aggregated together by gravity. However, DNA, like an elegant ballroom dancer, found a way to stay together amidst all the vicious elements and conditions. For an encore, it found a way to reproduce itself, something that no other organic molecule has done before or since. The sequence of nucleotide molecules in DNA is arranged so that a template can be made, called RNA, which then runs off a ribosome, resulting in proteins that then go to specific places and do specific things that are necessary for life to exist. The sole purpose of a gene is to produce a protein. Simple. The proteins are the messengers, mediators and emissaries that effect a change somewhere else. They might go to the outer membrane of a cell and plug into it, causing a reaction where the cell produces a particular chemical that in turn performs the ultimate desired effect of the gene. There are thousands of genes in an individual and the higher up the species tree you go, the more genes. The invention of DNA happened in one cell a long time ago and all life has probably come from that single successful cell — unless there was a time and place where molecules were having a hyperkinetic party, trying all kinds of ways of preserving collections of reproducible molecules into cells.

One of two events probably happened:

1. DNA was bound and determined to make life happen.
2. Life *started* as proteins or other mediators of metabolism and then found better and better ways of consistently reproducing itself, finally ending up with DNA as the memory storage.

Science might offer more explanations for life coming into being, but these two polarized possibilities are probably at the top of the list. Both inherently contain huge questions as to how the process started and have to include the obvious: what was the driving force for molecules to aggregate to become life?

For DNA to come into existence first, it would have to foreknow that its ultimate purpose was to be the bastion of knowledge of its species, because in order for it to accurately reproduce itself through the process of first making an RNA template, then the RNA making the proteins that then guided all the metabolic work, it would have to know what the metabolic work would be. Pretty weird for that to happen, period, let alone for it to happen randomly.

The other option is that some primitive metabolism arose, that later became more efficient at preserving its species' information (and all the species would be single-celled ones at that time). Some molecular functions might have been occurring that the collection of molecules 'wanted' to *keep occurring*. In biochemical terms, that means that chemical reactions in consistent sequences existed, through positive and negative quantum relationships, and that the molecules somehow preferred these to all the other random chemical relationships. Maybe there were simple little molecules at first, but for them to get bigger would require some protection against the chaotic and harsh outside world and so the collection of molecules (not quite life yet) would have to fabricate an outer wall to keep the organized things *in* and the random garbage *out*. Today, this is a fatty cell membrane, so back then it might have been something similar. In the metabolic progress of things, the simple molecules that were busy going here and there within the confines of a cell membrane were becoming life. At some point within this safe milieu, chemical reactions came into being that allowed for the preservation of the protein mediators' composition by way of the more hardy RNA. After this, the next step would be where RNA made a reverse template of itself — like a memory in a hard drive — which would be a strand of DNA that would then have to find its exact mate, nucleotide for nucleotide, in order to make the second strand of DNA in order to complete the double helix.

By now, you might have the same question in mind for both these options: who or what is at the helm guiding all of these impossibly complex ideologies to be played out at the quantum level? Why should molecules go to all this effort to make life?

Why would a cluster of chemicals have a need or a drive to assemble together into some reproducible form? A star simply contends with the balance of explosive heat expanding it versus gravity of molecules collapsing it. Rock is just cooled star. A mud puddle is dissolved rock in water. These are all collections of matter and energy, in phases that follow predictability, and none show any pattern of needing to reproduce themselves.

The other huge question, that mirrors inquiries into existence, is: what is the business of these early little molecules going here and there inside a cell-to-be? One minute there was a puddle of muck containing all kinds of chemicals and the next minute some molecules are arranging themselves so that some order is attained. They were specializing in their activities so that the whole forum of the primitive cell could exist *and* reproduce itself. This is not just chemistry but now biology, not chemical interplay simply due to all kinds of reactions but organization into living reproducibility. This is a major gateway.

The question can be split into two:

1. What was driving the chemicals to create intracellular organization?
2. Once it had attained this feat of making a 'cell', what was the need or driving force for it to exactly reproduce itself?

It's as though chemicals really, really, *really* wanted to make a cell and, after attempting millions of ways of doing it and finally succeeding, that cell had the intelligence to remember and to preserve energy and time by going for the more expedient process of patenting its discovery instead of randomly trying all over again the next day.

Science would offer the usual promissory note that it will one day find out how this gateway came about and tell us. Some members of the God-believing group would be comfortable resting in the belief that God is the one who has given things a nudge. Along the way in this inquiry, we will come to similar crossroads, as we narrow the questions finer and finer at the tiny quantum level or at the enormous astronomical level, where some people will rest in the conclusion that God is the force that we've been looking for. At any step, from the beginning of cultured time to the present moment of knowledge, there is a choice about God — believe or not. When we've been here for another 10,000 years and control all the energy of our galaxy, the choice will be the same. For anyone who believes in God, the question of how God has made the universe and everything in it is interesting, but that question is apart from the relationship we have with God. The scientific realm explains *how* the universe works and the spiritual realm helps us to know *why* we live.

So life became the new kid on the block, never before seen on Earth but now working hard to transform the surface chemicals into transient forms of matter and energy (living beings) that are capable of exactly (to the casual eye) reproducing themselves — whales and fish, birds and reptiles, bacteria and mammalians, fungi and viruses, trees, shrubs, grasses of all kinds. If inanimate things like rock and water had a viewpoint, it would be that their domain was being totally invaded by a parasite that morphs their substance into something new and non-rock: life.

Reproduction

Mitosis

Dividing into two identical twins has been the status quo for single-celled reproduction and for multicellular organisms' growth. It's the dominant form of reproduction amongst the microbes of Archaea and Bacteria. There's no sex involved — just the doubling of everything inside the single cell, then pinching off the enclosing membrane so that the distribution of property is equitable and two very similar cells result. This process worked fine for three billion years when microbes were the only beings on Earth. The populations of cells were essentially clones of the mother cell because all that they were doing was the same process over and over again: divide into two. If circumstances changed — temperature, nutrients, ambient gasses, etc. — that did not support a particular line of clone cells, their lack of genetic variability would mean extinction of the whole line. Conversely, if the environmental conditions favoured a line of cells, they could reproduce dramatically, increasing their numbers exponentially. A while later, when things changed again, this successful line could be wiped out. *So nature used high numbers to make this system work.* There was *some* variability between cells due to minor genetic changes and from trade of genetic material between cells.

Meiosis

Then another gateway came along — in order to attain multicellularity and specialization *and* survive, there had to be more genetic variability in the line of cells. This led to species surviving longer because within their genetic group would always be some individuals who were slightly different and so slightly better or worse prepared for a change in circumstances. Sex was born.

As with the other gateways, this one begs the inquiry as to what the driving force was for individuals to get bigger or multicellular or to specialize. Some single-celled species can aggregate into large colonies of cells — within that colony can arise specialists and several different kinds of cells that perform specific functions that benefit the group. We can see that those two functions, sticking together and specializing, result in better survival for the whole colony. So the result makes sense. But what do single cells know of sense? They only know live or die. Somewhere in this transition to multicellular, there had to be the experience that cooperation meant survival, and for that to be 'remembered', the memory would be in the DNA.

So the final question to explain this gateway would be: how was the experience of chemically mediated biological success of single cells aggregating together transposed into the memory of DNA?

While mitosis simply creates clones with very similar genetic makeup, meiosis creates variability. The members of entire kingdoms, including the plant, animal and insect kingdoms, have reproductive organs that specialize in creating one half of the genetic makeup of the next generation. When mating occurs, each offspring ends up with two sets of chromosomes, one from each parent. In this process, the functional, uncoiled DNA is condensed into chromosomes, and the pairs of chromosomes can physically bump, touch and cross over each other, exchanging genetic material so that the resultant ones have a mix of ancestry in them. Then the process creates the sperm/egg for animals or pollen/spore for plants that has half of the genetic material that an individual of that species possesses. When mating occurs, one of these cells finds a partner from another individual of the same species, and the two combine to form a full chromosome number again. So any one individual stirs up its ancestry in its reproductive cells to mix with another individual who has done the same. This mixing provides offspring with lots of genetic variability, unlike mitosis.

So, sex increases the chances of a species surviving. It creates slow genetic drift, allowing for a broader range of types within a species so that when the big test comes along, like a virus, a competitor species or an asteroid, there are increased chances that some members of the species will have the equipment to survive.

Happy Dirt

To summarize up to this point, we have molecules that had a force push them into organization to become cells. Then when that was accomplished, there was another urge to maintain that creation by cellular division. Cells started conglomerating and somehow remembered that this clustering could mean longer life and species survival. To get bigger, into multicellular organisms, a new process of reproduction came into being: meiosis. You wonder if somehow it is the same driving force throughout, from molecule to cell to multicell to multicellular organism, or if each gateway required a new type of motivator. In any case, we get the picture that atoms wanted to be alive. Why? Planet Earth is about halfway through its lifespan and did just fine as water, rock and dirt. Mars and Mercury are pretty well just rocks and dirt, and are they unhappy?

Life was never content with what it was, there were always new forms of life coming into being — plants, fish, reptiles, birds, mammalians. And in the mammalians, there were hominids of all kinds, and eventually us, thinking about things. From the beginning of time, everything in our world has been driving towards this moment when you and I wonder what the force is that makes all this happen. You can break the questions down into smaller and smaller queries or lump them together into bigger and bigger theories, but they all point to the same question — *why?*

Molecules Affect Our Behaviour

To glimpse into the depths of biochemistry in another way, we could look at hormones. A hormone is simply a chemical that has its effect in a place distant from where it was made. That distance could be very tiny, as in the case of a nerve ending, such as where a nerve meets a muscle. It's *right there*, but there is a tiny gap that a hormone has to go across to do its thing and make the muscle work.

An example would be a hormone, abbreviated ACTH, that originates in the pituitary, at the base of the brain. ACTH is secreted into the blood and can touch all kinds of cells in the body — your toes, your liver, your heart — but only the adrenal glands have a place for it to park. The effect is that it tells the adrenal glands to produce another hormone called cortisone that in turn releases into the blood and has its effect all over the body.

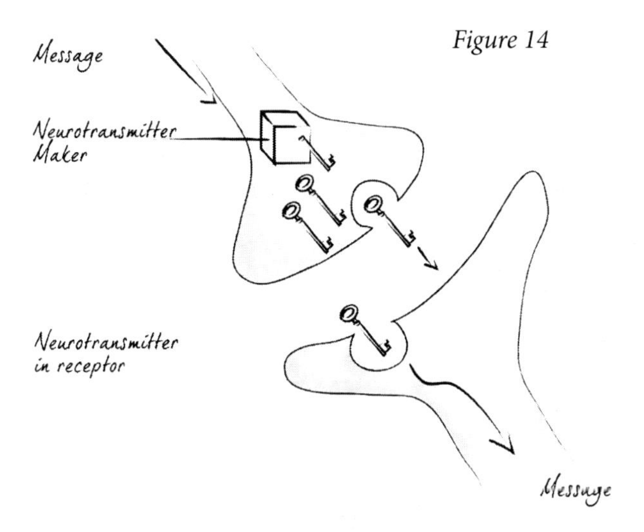

Figure 14

A nerve ending

This shows an ingeniousness of development where, as bodies got bigger, the distance between the maker of a hormone and its target organ got farther apart, and the body didn't necessarily need a nerve connecting the two, so instead had the target organ develop specific and unique receptors to accept the hormone. Thus, instead of nerves or physical connections branching out from the base of the brain and connecting to all the target organs, it went wireless, using the bloodstream as the way to send hormones to their target cells. As you can imagine, having all the wires made one more susceptible to having them broken by the trauma of accident or by physical attacks, and so they weren't successful. Wireless was the better model very early on in evolution, because all higher species — birds, reptiles, fish, mammals — have hormonal systems.

Of course there is always the possibility of the opposite occurring, where the hormones and their receptors were the first system,[3] followed by nerve endings stealing the idea. This would make sense because single cells have no nerves and muscles, but had metabolic machinery that required management over a distance. It was microns of tiny distances from one end of the cell to the other, but still some kind of a separation from the source of molecule production to its place of action.

There's always a second blasted theory.

This adaptation to continue using a hormonal system within the challenges of the enlarging physical body could be another gateway in the development of life. The other possibility, that cells created a hormonal system that could still be the model system for a hundred-trillion-cell multicellular organism, is suggestive of premeditation or luck or uniformly applicable principles. The body could have stayed at a certain size, but like everything else, from atoms to cells to animals, there was a drive to develop, to get bigger and more complex. Cells (or molecules) were cooperating to respond to physical challenges. Just as there was no apparent reason for muddy water to become cells, so there is no easy explanation for multicellular getting bigger and more complex.

Animal life is not the only realm that shows ingenuity of design. A big old tree that became a log without its bark, lying on the beach or in the forest, shows that the trunk makes an even, longitudinal twist. This didn't occur after death but reveals how the tree grew and its responses to increasing pressures put on it while making a vertical support structure with horizontal branches. The cells that made up the tree laid down woody substances in the spiral formation. In doing so, they had to communicate with each other that the bigger structure of which they were a small part needed the spiral shape. There had to be some kind of barometer receptor that detected pressures. The living cells in trees in most of the northern hemisphere are just under the bark, all around the casing of the tree, so the active

3. This was brought to my attention by a science student over beers at the Swans pub.

part is a kind of tube. We see from tubular, hollow structures that the spiral or helix is the strongest configuration, such as sheet steel wound to make a cylinder like a ski-lift tower. Another example is the cardboard tube that toilet paper or paper towels are wrapped around.

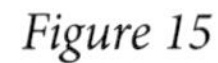
Figure 15

Log with long, gradual twist in it

Physics agrees with the trees and the manmade cylinders that the spiral is the strongest way to go. The trees agree with the animals that more is better — more cells together, more complexity in design, bigger size and greater surviving numbers. Everybody wants to cover the Earth with their species.

Rejuvenation and Cancer

Cancer is a result of a chromosomal aberration, so it is a mistake in one single cell at the molecular level. Chemical bonds within the DNA of the chromosome are broken from some insult, usually a repeated insult. Many singular insults are repaired by intracellular mechanisms, or the cell ceases to function. If it weren't for the unexplained drive discussed earlier in this chapter which pushes everything to be *more*, in the event that a bit of genetic material in the chromosome got irreparably broken, it would simply mean the death of the cell. No problem, there's lots of other cells. However, if the breakage occurs at a place in the chromosomes that has to do with control of cell reproduction, then you end up with a single cell that becomes a runaway lunatic, cloning himself without listening to his neighbouring cells. It's as though the reproductive part of the cell got its accelerator stuck to the floor or lost its brakes. In either case, the result is rampant reproduction.

At the margin of a cut in our skin, cells are busy responding to the need to heal by reproducing themselves. At their level, it is cellular reproduction, but from the point of view of the whole individual, it is the organ of skin regenerating or healing. The same thing occurs in most other tissues, such as in the liver, the heart, the bone. The organ that has been injured repairs itself by the process of the cells near the wound making more cells. This is a miniature version of the growth phenomenon that occurs in all living beings between birth and maturity. We grow to a point then stop; if we didn't stop growing, we wouldn't live long.

We see, then, that growth and healing are part of the mechanics of cells and organs, and that cancer is a breakdown in the regulation of that process. So when people say that we want to rid society of cancer, it isn't that we want to throw out growth and healing; rather, we want to detect early when the regulation of those things has gone awry and make the repairs before the insane cells take on a life of their own.

These few examples show us that much of life is managed at the molecular level. It reminds us of the mystery around the events when atoms decided to group into molecules that became living cells. The sustainability of life is as wondrous as the creation of it.

4

Life in General

The Big Picture

Different forms of life have held sway for eons.

First, there were single-celled organisms like bacteria, then plants, then animals — with large reptiles known as dinosaurs reigning for 150 million years.

While microbes and plants still dominate the planet, animals have had some impact in changing the surface of the planet. Of the animal kingdom, the most recent addition has been mammalians, and from this group there has relatively recently arisen a genus of two-legged beings called hominids. Several species of hominids have come and gone, but one remains that is having an enormous effect on the planet — by the name of *Homo sapiens* or Sapiens.

This species acts like a virus, gobbling up resources and secreting wastes in exponential fashion, on its way to being defined as a poor parasite — one that kills its host, which in this case is the Earth.

This species has the capability for thought and conceptualization, which allows it to be conscious of space and time. It has the capability for cooperation and inquiry into the mysteries of the universe but spends much of its time focused on fighting and sex, both of which are animal pastimes, as well as on alternate realities through drugs.[4] Scientific knowledge predicts that such a species will go extinct. As the members of this species left the animal world and experienced more and more consciousness, they also found ways to augment the evolutionary process where energy spent on simple survival evolved to power over others and then to runaway selfishness, control and greed. There is no question that millions of Sapiens focus on the creation of positive projects, on higher aspirations, on cooperation and fairness, but when these attributes are pitted against the more negative side of our evolution, our destructive aspect always seems to tear us down. Witness that in the past 2,000 years, every empire proved unsustainable and collapsed, and that the empires have become progressively more destructive to other people and to the Earth.

The reason to address this phenomenon and trend is not to bash Sapiens but to get our species to focus on the possibilities. Will the negatives of evolution win out and decimate our numbers, perhaps driving us to extinction, or can we choose to rise above these forces and be part of the process of evolving into a new, more cooperative species?

In any case, from our success or our mess might arise another *Homo* species that can do what we could not, just as we attained more accomplishments that our forebearer *Homo erectus* and cousin *Homo neanderthalis* could not. Other possibilities will be seen later in the book.

Parallel Worlds

I had a marvelous experience one day while scuba diving. A buddy, Graham, and I were following the breakwater on the shore in Victoria as it descends along the gradually deepening ocean floor, when out of the black depths came an enormous lingcod. They're long

4. This is a description of the top three worldwide commodities — in order: arms, drugs and sex — most of which will never appear in any Gross Domestic Product figures.

and skinny, like an eel, with a brownish body and big black spots. He came within 10 metres of us, checked us out, and then swam back down into his backyard. It hit me then and there that if you were to change the medium from H_2O to O_2 — a liquid to a gas — that this being I had just encountered could have been a dog. I experienced a glimpse into a different paradigm on the evolutionary tree.

Imagine the possibility of an entire evolutionary tree based on another chemical system. All of life on Earth is based on water, H_2O, as the liquid via which all other activities of life take place. In the basic building block of life, the biological cell, all activities that support the cell — namely, absorption of nutrients and oxygen, metabolism and excretion — occur in the chemical vat we call water. Without water, there's no means for all the chemicals of life to freely bump into each other and have their interactions. Water is copious and relatively unaffected by all the goings-on around it, yet very able to contribute or take away a molecule here and there when necessary.

It's theoretically possible for life to exist on the basis of some chemical other than water, such as NH_3 which is commonly known as ammonia. If that were possible, then planets like Jupiter that we might surmise couldn't possibly have life could indeed have life. However, as it appears that all of the basic building block molecules in our bodies and in life in general occur in the same proportions as they do in the universe, you wonder if a non-water-based life system would hold these same proportions. Then again, no one said that it had to.

The partner to water in the building of life is temperature. Compared to the rest of the universe, life requires a very narrow band of temperature to exist. Molecules can exist at very high and very low temperatures, which would describe their energy levels, but outside life's comfort zone the molecules do not aggregate into lifeforms.

You have probably noticed that when you look at a full mountain from floor to peak that the vegetation mimics the same gradations that are seen from space looking at the continental vegetation as you go towards the poles from the equator. There is water everywhere; however, when it's too cold, it is either frozen or the molecular actions are so slowed that metabolic functions cease. In this view, for the dance of life to exist, water and temperature have to marry.

Figure 16

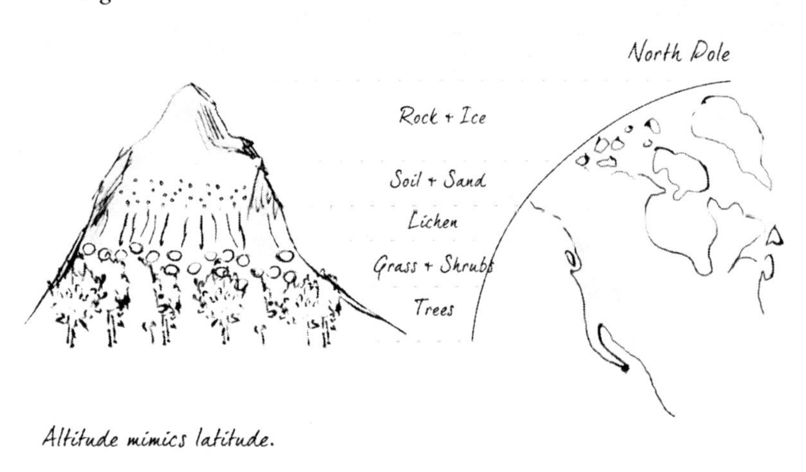

Altitude mimics latitude.

Zones of life, based on water and temperature

When you think of zones of life and the factors that affect them, there is the concept of zones around the Sun, or any star, where planets too close are cooked and the ones too far are frozen useless. It would appear that there are paradigms upon paradigms of zones: distance from the Sun, distance from the poles, distance up the mountain. When we think the other way, out into the universe, are there spatial advantages with respect to the Solar System's position within the galaxy, our galaxy's position within our galaxy cluster, our cluster within the supercluster and all of that in relationship to the entire universe?

Irrepressible Life

All of the life processes in a cell mentioned above occur in every cell, regardless of where that cell exists on Earth. So the single-celled species like bacteria, algae and amoeba do it. Every multicellular member of the evolutionary tree, whether plant or animal, is composed of these individual building blocks. You can get fantastic levels of specialization whereby, for example, the liver cell performs specific functions and is not capable of seeing like a retinal cell or of physical action like a muscle cell. Whenever cells of an individual come together, they have this eerie ability to agree that their common goal is to survive and reproduce their larger organism. Why? We might not yet comprehend why life came into existence in the first place but here, with life all around us and with our advanced understanding of how life works, we are still left empty to the question: what is the driving force to allow life to *continue*?

There Is a Driving Force

We don't know what the driving force is for your life to be assembled and thrust into the world. Perhaps it's to have your entire body do what your first cell did: become two. After all, if we presume all life on Earth came from one cell originally, then the human that arises from one cell, a fertilized egg, also has enormous potential that becomes manifest in the sum total of all the cells' individual striving for life. So you continued to grow until you were capable of reproduction, somewhere around the early teen years. Still, if people flowered (reproduced) and died, the terribly dependent offspring would die too, so for the system to work, you had to live to be that age twice, to allow you to care for your offspring until they reached reproducible age.

So by the time you reach 30, the evolutionary system of reproduction would be satisfied. We still don't see the reason for you, and now your offspring, to be alive, so maybe the answer is found in the concept that the whole species needs to survive. Still, our species could not survive without other plants and animals on the planet, so there must be a deeper drive, propelling *all life together* to a common survival goal. We don't know what that is. It could be that Life has an ultimate purpose apart from Sapiens, and that we're just a blink-of-the-eye sideshow in the larger geological timeframe. If the ultimate purpose is for Life to discover and harness the secrets of the universe, then Sapiens has a chance to contribute to that. It sure looks like all of life is expanding in complexity, as illustrated by hominid history, where a barren Earth nurtured life into being and now we are here, talking about it. If Sapiens is truly the leading spearpoint of knowledge and ability to unravel the mysteries of the universe, then we are Life's representative, which carries an incredible responsibility.

Inherent somewhere deep in Life, before even DNA, is a drive that we haven't identified yet, one that pushes for bigger individuals, made of more cells, and more individuals, making a larger population of organisms. Everybody does it, from single-celled organisms to humankind. There is no internal regulator to restrict increases in species' numbers and, without external regulators like restriction of nutrients and factors that decrease birthrates and decrease in numbers by disease, predation or disaster, then the population could reproduce without limit until the Earth is consumed. No species has ever consumed Earth, and some force has always come along to re-establish global equilibrium. Can we learn something from this?

Will Animal or Choice Win?

Dinosaurs were around for 150 million years. Neanderthal was king over us for thousands of years. We've been the dominant species for about 20,000 years, but our impact on the Earth started only about 5,000 years ago, with most of the advancement/damage incurred in the last 100 years. We're rapidly reaching that point where we will have consumed all the resources. Like the fermentation of wine that finishes when the yeast is killed by its own production of alcohol, we are going to drown in our own wastes (credit to Kurt Vonnegut for that analogy).

The problem is that our consumption of resources, including territory, means that there are less for all the other species, and the wastes that we produce are toxic to them. So the extinction rate is climbing fast, simply because Sapiens is successfully reproducing. Clearly, then, there is more to the Life on Earth Project than survival and successful reproduction, and so our response to our primordial fears — to gather power, importance, food, money and property — hasn't taken our species to a constructive place where we're closer to balance, knowledge and happiness.

The dinosaurs had a successful way of being, but that was largely because the Earth was warmer and wetter, allowing for lots more vegetation, to sustain probably five times more species of herbivores than exist today. Those herbivores were the walking hamburgers for the carnivores. It was a hyperkinetic eating party. Well, that party ended and Life had just partied and not gotten anywhere. It would seem pretty safe to say that the dinosaurs didn't stumble across the purpose of life, unless the purpose is indeed to simply be part of the basic stuff — to eat, drink, move your bowels and reproduce.

Somehow, although that simplicity might be a good way of being, we know in our hearts that there is more going on than that. The universe is so big and we've only started scratching the surface of what exists, let alone how it works. There are a small number of brilliant people inquiring into the way the universe works, trying to unravel the mysteries, calling us to a higher purpose, and God-believers say that He is calling us too, but our animal cores often mess it up. We can understand that humankind is one big organism and, as this organism grows, it appears that the polarities between good and bad are growing too — genius and depravity expand in balance. Consciousness empowers and destroys. Still, the genetics dictate that Sapiens constantly looks to survival and lives in fear of possibly not surviving. This frustrates our ultimate potential by physically overpowering the intangible. How can joy, love and God survive in us when we're stuck in fear — trading guns, drugs and sex instead of trading knowledge and talent?

Which Path Is Sapiens On?

In the book *The Teachings of Don Juan: A Yaqui Way of Knowledge*, author Carlos Castaneda cites Don Juan as saying:[5]

> Anything is one of a million paths. Therefore you must always keep in mind that a path is only a path; if you feel you should not follow it, you must not stay with it under any conditions. To have such clarity you must lead a disciplined life. Only then will you know that any path is only a path, and there is no affront, to oneself or to others, in dropping it if that is what your heart tells you to do. But your decision to keep on the path or to leave it must be free of fear and ambition, I warn you. Look at every path closely and deliberately....
>
> ...Does this path have a heart? All paths are the same: they lead nowhere....Does this path have a heart? If it does, the path is good; if it doesn't, it is of no use. Both paths lead nowhere; but one has a heart, the other doesn't. One makes for a joyful journey; as long as you follow it, you are one with it. The other will make you curse your life. One makes you strong; the other weakens you.

When we look at our society in general and our schooling specifically, we see that most of our training is intellectual,

5. I found this scribbled on a piece of paper in a cigar box, a note-to-self from forty years ago.

exercising the mind. There is some exercise of the body, but by the end of high school most can opt out of participation in that. There is very little education, experience sharing or training in emotions or the Heart. There is usually none in the department of the Soul.

We see now that as the electronic age progresses, more and more people, especially youth, are plugged into electronic communication. Look to your left and right at the next stoplight and see that the passenger is looking down into his or her lap, intently. They might be playing an electronic game but are probably texting on a phone device or using a software application via wireless Internet connection. (That sentence could not have existed before this century.) This is one step more cerebral and one step less emotional because the transfer of information is all text with no visual exchange of facial expression nor auditory exchange of nuances of tone. This will lead to more brain development for text information exchange, with atrophy of other parts of the brain, just as the written word allowed our verbal memory to shrink. This isn't necessarily a bad thing — it might be the next evolutionary step to take Sapiens to new heights of development.

Thresholds

It may seem obvious to say it, but every system has a containment within which it works and beyond which the system breaks down. Sometimes at that breaking point, a new creation or a different system comes into being. Notice how ants only get so big and no bigger. When a species gets low enough in numbers, there is no recovery and they slide into extinction. We are miniaturizing silicone chip technology to the point where the circuits will be atoms wide, which won't be able to carry the information any longer.

DNA is composed of just four nucleotide molecules, arranged to 64 'words', or codes, for RNA to transcribe into proteins, which are the effectors of genes, the messenger boys who do all the running around to maintain a life. If there was one less nucleotide, then life wouldn't be possible. If there was one more, it might create confusion and mistakes; again, no life.

There were no quarks in the universe and then there were zillions — the universe was born.

There was no life on Earth and then there were cells with DNA.

There have been thresholds that resulted in creation, and there are strict guidelines that hold thresholds in abeyance in order for life and the universe to continue to exist.

Reserve Pool

All of the many different aspects of the natural world, from atomic to astronomic, arise from a reserve pool of building blocks. There is usually tremendous wastage — portions of the creation that are never used or that are discarded as useless for the project.

At the astronomic level, there are untold numbers of atoms that have aggregated by gravity into stars and planets, but it's not a tidy universe. There are entire galaxies that are colliding, destroying planetary systems like our Solar System, with new creation of celestial bodies resulting. There are random wild cards like comets streaking through a galaxy. If this Solar System has an asteroid belt composed of zillions of chunks of lifeless rock, you can bet that there are equivalent unorganized planetary messes on far grander scales out there in the universe. Once we've contemplated all that, we have to remember that what we study and see is only 5 percent of the universe. Much of the mass that we can't see is dark matter, an example of which is the black hole, an incredibly collapsed aggregation of atoms, crushed together by gravity, without electrons to space them apart.

We intuitively see that in all this astronomic marvel, there are tremendous forms of reserve pools, in this case of matter and energy.

When we zoom in to our Solar System, we see that Earth is a small part of it but that, without all the other parts, the Earth could not sustain life.

On our planet there are many forms of life, but the predominant form is single-celled, as we've seen, where Archaea and Bacteria form half of the biomass. This is how life started (as far as we know) and so even now, a billion years into complex life, there is still the reserve pool of life, ready to start complexities all over again. This brings to mind the notion that single-celled organisms broke out of that pattern at one point and became multicellular. Why have they not repeated that feat in the last billion years? Or have they and we haven't detected it? It's as though there was a gateway at a point in time, after which the gate never opened again. It's possible that simple cells and DNA would be eaten if they now tried to come into existence, so while the opportunity to create new life might still be with us, the gateway is closed due to the already-existing life. Are there other gateways that are no longer effective due to the existence of that gateway's previous progeny? Perhaps gateways open all the time, but without the right conditions to receive their resultant effects, it is insignificant. As sure as nature abhors a vacuum, all of nature is eternally poking around, attempting to penetrate deeper into every aspect of existence, to expand, to create, to be more than it was yesterday. Is not Sapiens obeying that ordinance, riding the wild vehicle of evolution?

5

Stars and Planets

What Is Our Destiny?

Three of us buddies were bombing around the countryside on our motorbikes when we stopped in at a funky little café for breakfast. Both of my companions have inquiring minds and lots of ability — they're the types that could rebuild a bike at the side of the road, and those who ride know exactly what I mean. Someone brought up the news that there was an atomic facility in Europe that was in the process of producing a small black hole, and Roger said immediately in a mock-panic yet level tone, "Oh that's not good!" Gary laughed, understanding what Roger meant, and said, "Yeah, what if that's our purpose and destiny in the universe — to get to this stage of development, to create a black hole that sucks in the entire universe, from which another Big Bang occurs and the planets and life start all over again?"

Things Add Up

If Earth is spinning around the Sun at 30 kilometres a second, and the Sun is swirling around the Milky Way galaxy at 217 kilometres a second, there must be times when the two speeds compound, like on the Tilt-a-Whirl ride at the fairgrounds, when one body is heading away from the centre and the smaller body within that centre is also heading out, and so for a moment the smaller body Earth, is *really* winging out around the orbit.

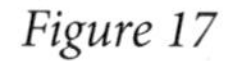

Figure 17

Tilt-a-Whirl ride

And if our Local Group[6] of galaxies is moving around the centre of the Virgo supercluster at 600 km/sec, then there'll be times when Earth is moving at 30 + 217 + 600 or 847 km/sec, compounding the speeds. The entire Virgo supercluster is moving around other superclusters and no doubt, if infinity exists, there are groups upon groups, in larger and larger scale, moving in orbits around each other, each adding their speed

6. Astronomers group celestial bodies by their spatial relationships, so the Milky Way Galaxy is 'circling' around a centrepoint with other galaxies, and those galaxies, called the Local Group, act as one large cluster that moves with other large clusters in an enormous pattern called the Virgo supercluster.

to Earth's final speed, based on the farthest defined centre. The speed of light is 299,792 km/sec (187,000 miles/sec). What happens when, relative to some far-off centre, planet Earth, on spinning outward on its elliptical orbit, attains this speed? Maybe nothing, because the reference point is so far away, even for light. Maybe the speed would be most important on a smaller scale, like the speed of an electron going around a nucleus in the tip of a tree on the top of a mountain on Earth. When that electron reaches the speed of light relative to the distant reference point, perhaps a paradigm will be leapt. Maybe it leaps all the time and we don't know it yet. Maybe that envelope around the whole body of stars and planets (the universe) defines the end of the 'Universe' and the beginning of another universe, and the universes are moving around each other in different planes or dimensions. Are the last vestiges of the previous Big Bang visible on the outermost reaches of our universe? Or has another Big Bang occurred far away out there and its leading edge celestial bodies are just approaching us, in the same way that galaxies collide?

From Quarks to Planets

Whenever some new facet of our physical world is discovered, it appears as though it was there all along and we just needed someone to point it out to us. Part of the process seems to be that our brains are not able to understand or incorporate what is there until we take several runs at it. The factor that inhibits open-minded new learning is fear. One fear is based on inertia: we want life to get better, but stay the same. In other words, we fear losing what we've got, but want more. The other fear is of insanity. We have constitutions and conventions, mostly fabricated, that describe our worldly reality and we need to work within that realm or our thoughts and societies become explosive and chaotic, destroying any progress in conceptualization. We send brilliant minds to the outer fringe to explore and return to us in our safety, and report what is out there. We learn in small steps, expand our minds and eventually we all go to the fringe, like a cautious herd of cattle investigating a new thing in the corner of the pasture. But by then, the fringe has moved further outward into the unknown.

Other Earths

This is the current astronomy flavour of the month, that there are other planets out there that are similar to Earth, with the automatic assumption that amongst them all there is bound to be life, and if there is life, then there could be intelligent life, like us.

Still, when we take a closer look at Earth, it has more than a few unique aspects that allowed life to come forth.

1. One of the theories is that, back when the planets were forming, Earth's moon, Luna, collided with Earth and, in the wrestle, Luna took away significant amounts of Earth's mantle, which allowed for:

- less metal for the moon, more metal on Earth, especially more metal close to the surface, available for use;

- the greater iron core of Earth to serve by magnetic field as a protector against radiation from the Sun's solar winds;

- leaving the Earth pear-shaped with an irregular surface. Had this not happened, the Earth would be more spheroid, like other planets, and so if water formed, it would uniformly cover the planet, removing the mix of oceans and continents that exist today.

2. Earth has the mineral composition that could break down to release oxygen.

3. The moon uses its gravitational pull to keep the Earth on an axis, otherwise the planet could tumble, which would create harsh climatic conditions.

4. The Earth is close enough to the Sun to keep water liquid and warm, but far enough not to get baked.

5. There are larger planets farther from the Sun — Jupiter and Saturn — orbiting around like bodyguards, using their gravities to fling possible stray celestial bodies from colliding with Earth. For many planetary systems, the bigger planets are closer to their star, and the smaller ones farther away, which wouldn't work as well for warmth and protection like our Solar System works.

6. The Earth is far enough from the centre of the galaxy to avoid the big black hole there.

7. The Earth has water, supposedly delivered by comets and other wandering celestial bodies.

8. Compared to other planets and their moons, Luna is pretty big relative to Earth and so its gravitational field has a big effect on the oceans, causing bigger tides to come in and go out, creating a gentle, repetitive mixing of elements, making an interface between water life and land life. If the water–land interface was static, there might not have been a transition from aquatic life

to terrestrial life. The tidal zone was a practice area for life preparing to leave the sea.

Figure 18

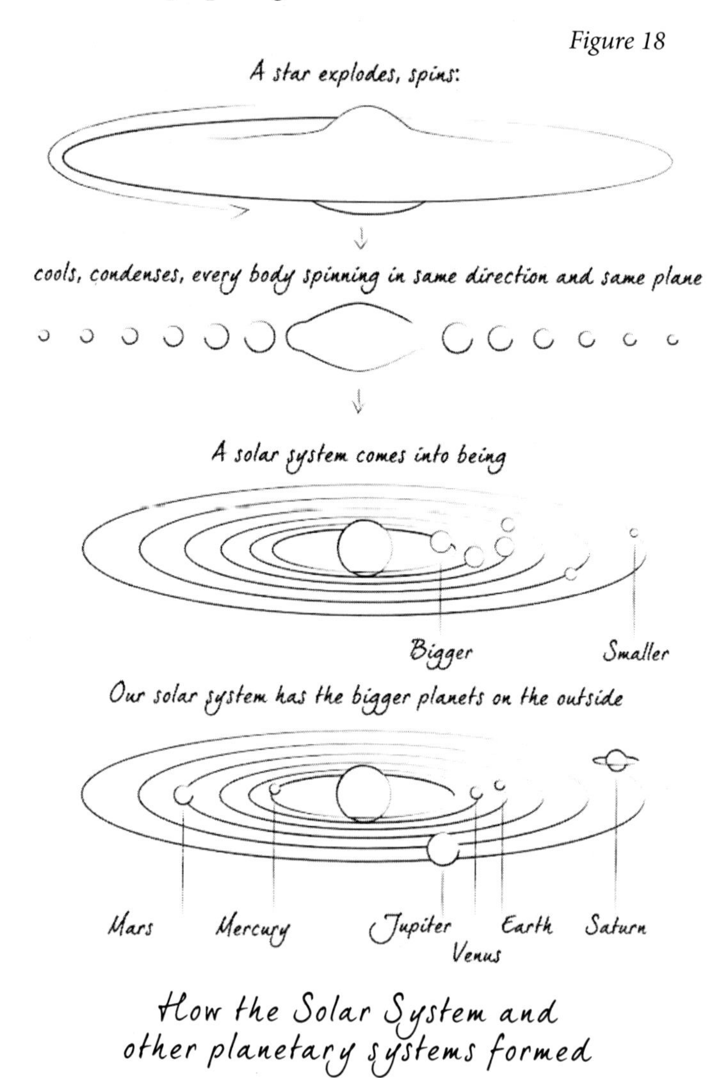

How the Solar System and other planetary systems formed

Current Summary

Some window opened for a moment and all the universe came into being. All or most of the other billions of tries at this failed and those universes ceased to be. In this universe, there are countless planets that orbit their stars, but some are too close and hot, and others are too far and cold. Many don't have a moon to stabilize them, or the right proportion of elements to allow for the generation of available oxygen and liquid water. Most don't have big brother planets like Jupiter circling around like a wolf, to scare off random intruders. So our home planet is unique — it might not be the only one out there that can support life, but it is a rarity to have so many variables come together in just the right combination.

The Earth Is Not Flat

Our Solar System was created from a fiery gas cloud that eventually started to rotate due to a balancing standoff between gravity of the particles and the heat expanding them apart. The particles would be drawn to the centre of the mass but pushed away by the heat, and each particle would shear off to one side, with the direction being random. The total net effect of all the particles would summate to a single direction, and once the mass of the early Solar System gas cloud started to turn, then all the particles followed. So everything was spinning around the centre in the same direction *and* in the same plane, like a giant discus or clam shell, and the outer belt cooled to form the planets. So when the Big Bang occurred, the initial heat and expansion was so explosive that the original body would have been a sphere, but once things started slowing down, wouldn't it behave like a galaxy, starting to spin one way and flattening out into a disc? Relative to the absolute nothingness of the non-universe, the Big Bang universe floating, spinning, expanding in that nothingness would have to be pretty flat.

Astronomers tell us it's not flat — when you add time to get the space-time continuum, they tell us, it's curved. What this ends up meaning, to us simple minds struggling to comprehend, is that the shortest distance between two points might not be a straight line but something completely different. It might be straight but our minds can't think outside of three dimensions in order to see that it is the dimensionality that wraps, curves or winds around the straight line. To get there, we will need not only tangential thinking to find the shortest path to the stars, but a propelling force beyond Model-T Ford rockets pushing a tin can through space.

Figure 19

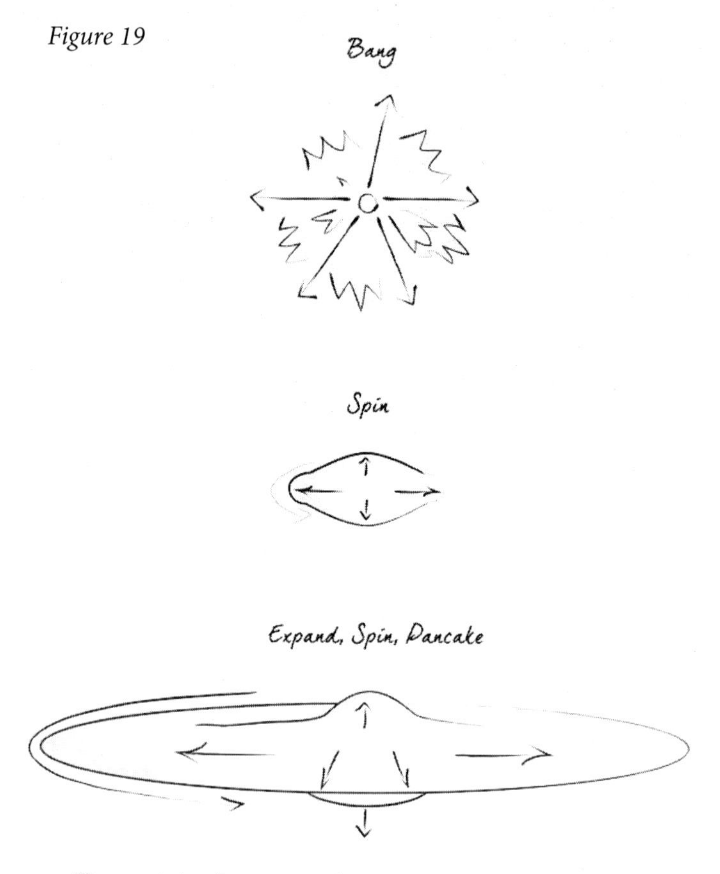

This is what galaxies and solar systems do, but not the universe?

From Big Bang to disc

Particle Evolution

While we're questioning everything, it seems like we're trying to break the universe down into its component parts in order to understand where all these little and enormously different particles came from — graviton, neutron, quark, meson, Higgs boson, etc. — and include them all in a grand theory. Could it somehow be that the equation is really simple if all the particles were basically the same in the beginning and that they have somehow evolved, like everything else? It would fit the 'simple and elegant' guideline for hypotheses if the particles that first came into being *were all the same*, and the incredibly high temperatures, zillions of degrees, led to the creation of further particles by the laws of physics.

Perhaps particles diversified because they had to, because if they didn't change in the way they did, the universe would've fallen apart like it did in the many other tries at big bangs that didn't work. If this could be proven, it still wouldn't answer why the first particle came into being, and what opened and closed that window for a zillionth of a second. Nor would it prove, if it took many tries to successfully make this universe, what the driving force was to keep trying.

The line of reasoning to back this thought is that everything came from nothing, all the energy of the universe balances to zero, and nature is conservative and efficient, so when matter first came into being it was likely all the same stuff. *If* it was quarks repeated by the zillions in a gigasecond, the temperatures, pressures, timeframes and spaces would be unbelievably exponentially big, which would be a forum to create changes in the quarks, rendering them into gravitons and electrons in the beginning, then mesons and neutrinos later.

Song of Physics

In Western music, there is a scale of notes: A, B, C, D, E, F and G — for simplicity's sake, let's forget flats, sharps and alternative scales. These are resonances or vibrations that reverberate through the air, travelling through the atmosphere to our ears where the vibrations are received, which sends a signal down the nerve to the brain where the message is interpreted.

We hear music. So there exists the orderly spectrum of notes, the medium for it to exist (air) and the equipment to receive and understand the original notes. When we look at the other spectra and energy forces of the universe, they are very orderly and consistent, they have a medium that they can exist within and there is a way to 'hear' their story. You have to wonder what other 'musics' might be out there, that we either haven't discovered or that we don't have the sensory ability to receive.

In addition, if each note from A to G was represented by a factor of the universe then applied in a systematic way, there could be 'songs of physics' that produce characteristics or phenomena of the universe. For example, if each note was a factor and you took another aspect of the universe, like an atom, and subjected it to these physics notes in a repeatable succession, like a song, you could effect a change or create something new. This would be "The Atom's Song".

a. magnetism
b. electricity
c. vacuum
d. heat
e. cosmic rays
f. light
g. gravity

Or, if you started with particles, and subjected them to certain energies and then played the melody, this would be "Energy's Song".

a. quark
b. electron
c. neutron
d. proton
e. graviton
f. meson
g. boson

Or substitute with different forms of energy such as parts of the electromagnetic spectrum, gravity or heat, and then play the melody. Or create a symphony where several different modalities are engaged at once. Perhaps the grand symphony would be the universe unfolding.

Time Travel: Danger, Danger

When you read about physicists' inquiry into time travel, they talk about wormholes, dimensions, etc. and that the energies necessary to create and sustain a portal would be all the energy of a galaxy. This all sounds pretty impossible for us for the near future. They also talk about going back in time, but little about returning again to the present time. Of course, this is all theoretical anyway.

However, it seems as though we are looking at time travel as a separate entity from the space-time continuum, which could be an assumption that isn't possible. A simple example would be for us to step into a time travel device that takes us back 50 years. We assume that we will be in the exact same spot, but without programming that in, you could end up in a *place* very different than you planned. The Earth is rotating around the Sun at 161,000 kilometres (100,000 miles) an hour, and our Solar System is at the edge of a galaxy, the Milky Way, that is spinning on an axis around itself. The universe is expanding. These three variables alone, if not taken into account in your time travel, could place you in interplanetary space a million miles from anything. Pretty scary stuff.

Still, you could take that notion and apply it to your advantage. It appears that it takes as much energy (currently rocket fuel) to get off the Earth and away from the atmosphere as it does to then travel a billion miles through space. So if you could take a small piece of a space station and transport it through time for just a few seconds into the past, but at the same 'location' in the universe, it'd be 100 miles in space, where the Earth *was* a few seconds ago. If this could be made accurate, you could repeat the process until you had a space station assembled from which you could launch rocket ships off into space. Then, if you could make the location (space) aspect of the equation a variable instead of a constant, you could pick not only the time but the location as well, which would make rocket ships obsolete, except maybe as a kind of hotel to stay in once you got to your location. By manipulating time and space, you could end up with parts and people scattered all over the space and time continuum, which would require consolidation and regrouping at some point. Perhaps the process will occur in steps, where you keep space a constant and vary time, then keep time a constant and vary space, and so on until everybody is where you want them to be.

Another aspect of time travel is that it is from the point of view of distinct timeframes, like big shoe boxes, where you can step out of one into another. Thus, you theoretically could go back or forward and meet yourself. Time and space are connected and it seems *improbable* that, by playing with time, you could suddenly have two of you. If it was possible that two of you could be in the same time then it wouldn't be in the same space. All the energy of the universe has to have a balanced net zero level, including the energy of the molecules that make a person, and attempting to have two of the same person in the same space and time would be up against tremendous resistance. So maybe you could be in a time bubble, with its own space, and looking at yourself, in another space, at the same time.

That's one issue.

Another issue is the concept that any body is a walking biological clock — the DNA is the watch master that sets the pace, slowly decreasing regular repair of wearing tissues and slowly decreasing elasticity and hydration to make a progressively more stiff and brittle, smaller and older individual. Everyone in any timeframe has the same DNA clock. Different species have different lifespans, but we are all guided by our set DNA.

To a fly that lives for a day, people live 365 days/year times 80 years = 29,500 days or 29,500 fly lifetimes. If a species on Earth had that many generations of longevity over us humans, given that a generation for people is about 20 years, that would be 20 x 29,500 = 590,000 years. So the fly, in our terms, thinks we live for 590,000 years. Isotopes, rocks and planets are described in this way.

If someone rockets into space, travelling at the speed of light, and 10 seconds later for him it is 100 years later on Earth, he would still be, by his DNA clock, 10 seconds older. Similarly, if he went through a wormhole, back in time, and somehow re-emerged the next day 100 years older, from his viewpoint he would've still had his 100-plus biological years of experience and worn-out body. The point here is that there is no way to avoid growing old by time travel.

Everything Is Relative

Now, to the practicalities of time travel — a list.

We are adapted to our time and space for temperature, humidity, oxygen, water and food sources, etc., but also for immunity. Our immune systems are picking up hundreds of antigenic signals daily and responding by making antibodies and programming white blood cells to gobble up the intruders, which are bacteria, viruses and parasites. These are always evolving too. If we time-travelled either forward or backward, we would be hit with an overload of unfamiliar intruders that would be different than today's, and we would probably succumb to one of them pretty quickly.

Jungle Bacteria

We were paragliding in Costa Rica when, on launch from a mountaintop, I didn't have enough speed or lift and went down into the jungle, left dangling, with my glider suspended over the jungle canopy. I had to shinny up trees, hanging on by my knees, gently lifting the glider off branches with a long stick. That all worked out okay, but beyond the exhaustion of working in the canopy for two hours in the humid tropical heat, in the next few days my right eye and a small cut on my right arm festered into bacterial mayhem. My immune system had never seen these microbes before. After a lifetime of small cuts that never became infected, a single visit to the tropical jungle canopy threatened to consume me.

To avoid being singled out as different or strange and promptly arrested by the authorities and thrown in jail, you'd have to match the clothing fashion of the day.

Language evolves quickly and as little as 500 years could make your own language unintelligible for you, going forward or backward in time.

Money changes too, so you'd have to take gold or precious metals with you to sell for cash in order to procure food and shelter — provided that those metals are worth anything in that future timeframe. For all you know, they could become sacred, designated for special use, or illegal in commoner's hands.

Going *back* in time, you'd have the advantage of approximately knowing the mindset and culture of the people, but going forward in time would be a huge unknown.

As you can imagine, if anyone found out that you were a time traveller, they would want that power. From your neighbour to the small-time businessman to the local authorities to the military to the government, you would be the single most valuable asset in existence. They would all want the ability to time travel to have power over others.

And that's the whole allure: *to be able to do something that others cannot* so that you can have an edge over them in some way, which is the old survival game again. If one person has an advantage over another, he might have a better ability to survive. As we've seen with Sapiens, the urge to survive has no regulator so when successful survival occurs, it is quickly followed by collecting and amassing stuff, land, people and power.

If the physics questions and the impracticalities aren't enough to dissuade you, watch *Back to the Future* again, where they needed to harness a lightning strike to gather enough energy to send Marty back to his time. If time travel takes a lot of energy, contained in controllable forms, you'd have to take some stored energy with you in order to get home again. The past wouldn't likely have the energy you need. We would all expect that the future *would*, but, with a chill up the spine, there might not be.

There Are No Road Signs in Space

Deep space is full of hazards. Only 5 percent of the universe is visible to us in the form of galaxies with their stars and planets, comets and nebulae. No one knows exactly where the other 95 percent is, but some of it is in the form of black holes, which are composed of mass that is fantastically compressed and therefore possess a phenomenal amount of gravity. Stars do this when gravity wins over the weakening explosions, and galaxies can have a big black hole at their centre, too. These would not be a good thing to fly near, as you would get sucked in and collapsed along with the rest of the molecules.

There are all kinds of remnant Big Bang stuff floating or flying around, like asteroids, that aren't part of any stable geography like a planet. They're not charted on a map, so you'd have to have an early warning system for collision avoidance, and probably a deflective force field. 'Deflective' because a force field alone might stop a monster chunk of rock from hitting your spaceship, but the force of impact would be transferred through the force field to your ship that created it, resulting in everyone inside being splattered against the windshield. A similar example is that a helmet can stop your skull from fracturing, but your brains might keep moving, getting compressed up against the impact side of the skull. Maybe if a force field had a shock absorber system it would work. If you had a means of blasting oncoming projectiles, you might just create a shower of rocks instead of one big one. Adding in the idea of rapid avoidance manoeuvres, there would have to be no life on board, as once again they'd get spattered against walls by the G-forces of going sideways. To be a good 'all-terrain' outer space vehicle, it might have to have all these aspects — early detection, manoeuvre avoidance, force field protection and ballistic disintegration of asteroids. Not having life on board would probably be best. A beer can would be better than the starship *Enterprise* from *Star Trek* — smaller is better all round (and this is part of what keeps spaceship planet Earth safe — it's tiny).

Apart from these road hazards, there are electromagnetic spectrum radiations — like cosmic rays, X-rays and gamma rays — that the Earth's metal ore core and its atmosphere protect us from. Many people know that one transatlantic flight exposes us to the equivalent of a single chest X-ray, because we're flying at 30,000 feet, where the atmosphere is thinner and less protective.

Even the simple lack of gravity in space eventually turns our bodies to feeble mush because there is no constant force stimulating the neuromuscular system, keeping everything in tone. So it would appear that heading off into space has more unknowns than sailing across an uncharted sea, where the weather, reefs, currents and wildlife were just variations of what we had already worked with. Planet Earth has been our playpen, and we'd better send robotic explorers off into space to collect data before we go there ourselves.

Nature Can Be Simple and Aggressive

One summer day when my daughters were young, about three and five years old, we were sitting at the patio table with its central umbrella shading us from the afternoon sun. Up inside the umbrella was a long, narrow fly with a striped body, flying around a bit aimlessly, unable to get on his way because he couldn't figure out to go down a few inches then fly away. Along came a big wasp. He too was inconvenienced by the umbrella and buzzed around inside for a while.

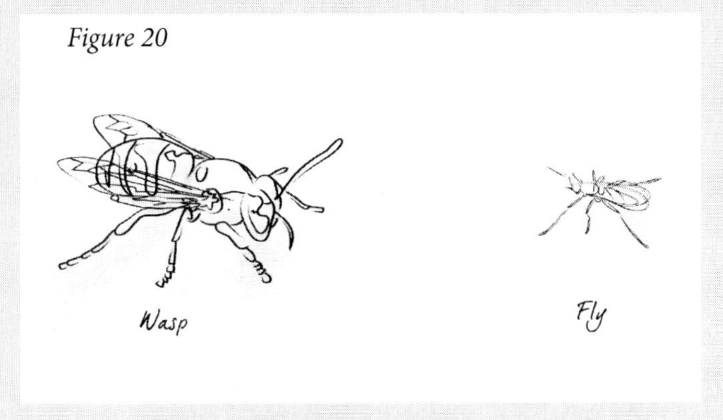

Wasp and fly

Then he hesitated, apparently seeing the fly, went straight for it, grabbed it and landed on the table. We watched in amazement as the wasp first snipped off the bug's head, then its wings, then legs, tucked the torso up under his body like a torpedo and headed off into the nearby woods, presumably to share the food with the nest.

This is a peek at evolution at work, and if it is working in the rest of the universe like it is working here at home, it would be good to know who is the wasp and who is the fly, *before* we meet any aliens.

Time, Space and Aliens

So there are dangers in both time and space travel. As we've seen from our species' history, the greatest danger to us has been ourselves. So outer space might harbour the greatest danger of all. If there are other intelligent lifeforms out there in the universe and they have evolved *as we have* and they are ahead of us in evolution, they could be many times more aggressive than us. We assume that, as we are the ones heading off looking for new horizons, we are the ones in control, but we could stumble across a culture that is more advanced than us. They might have a use for us or our planet, and take it. If they regarded us as tasty, we could end up like farm animals. If we had an asset that they could manipulate, we could end up slaves to procure it for them. If they needed our planet but not us, we could find that Eurocentrics with their history of colonizing the globe would be on the receiving end of the experience of being colonized. The entire planet of Sapiens could experience what thousands of species on Earth have experienced from our overwhelming presence — extinction by habitat displacement. Then of course, there is the ultimate disaster: extinction by extermination.

At this point, we could bring up God again and hope that the aliens know God and have compassion and mercy, which are intuitive and emotional traits that seem to go counter to the flow of evolution's pattern of self- and species preservation. The nonGod people might protest that they don't need God to have compassion and mercy — to which someone might say, "How do you know? If God has always *been*, then we have no space-time-spiritual continuum without Him to act as a control to see if humankind could have these traits without his presence."[7]

We know that without consciousness, we were animals. In that state, we had only natural drives, survival of the fittest and evolution. With the advent of consciousness came awareness of a larger sphere of being, including planning our time, taking care of more than just self and perceiving the spiritual world. We don't know the order of things, but if God *is* and He is outside of the constraints of time and space, and the creator of everything, then He must have preceded our awareness of Him. It is an idea to ponder: whether awareness of spirituality nurtured compassion and mercy, or whether somehow learning

7. Of course, the philosopher will point out the opposite view of a world without God and no 'control' to show what it'd be like *with* a God. There's always a second blasted view.

compassion and mercy led to spirituality. They are three separate variables that might have all come together, had sequential relationships, or little relationship at all. In the vat of human development, there might have been all those possibilities, occurring in different groups of people in different timeframes.

6

Heart/Mind/Soul/Spirit

HEART, MIND, SOUL: this is an artificial trichotomy, or three-way division, segmenting parts of who we are into smaller parcels in order to try to understand how our lives work. Some cultures split it up differently and not all cultures do this, but Eurocentric humankind is very familiar with it. There is no universally accepted definition of these concepts, but this chapter presents one stream of thought that might offer clarity.

Within this conversation, there is an overlap of the scientific and molecular with the more ethereal and spiritual and, while both areas of study are interesting, it is often the interface between them that holds the most excitement.

HEART describes the emotions, which are perceived and evoked through the brain but encompass more of us than that. It's quite probable that every emotion is preceded by a thought and so the brain initiates an emotion, but once the brain has processed the thoughts and evoked the emotion, messages go to other parts of the brain and to the body. Then the body is recruited to act out the response that the emotion dictates. You would think that emotions are the raw you, but that emotional part of the brain needs the guidance of the intellect before knowing how to proceed. So in summary, heart is really brain, emotions and body, in a continuum. We have probably given this part of our being the name 'heart' due to the heart's pounding in our chests when emotions are involved, as with fear, anger, joy and other passions.

MIND is the thinking and processing part of us, and includes, but is not limited to, the subconscious (and therefore dreams), vision and memory. If there is molecular or cellular intelligence, then all of our cells contribute to the sum total of our being. We know that parts of the body 'speak' to each other via touching each other, through the electrical depolarization of nerves, by hormonal communication and probably other processes we haven't discovered, so it's possible that there are non-neuronal cells that contribute to our thinking. The mind processes information, analyzes and interprets, and is the awareness reporter for all the cells in the body. The brain is just the spokesperson for the mind. 'Brain' is the anatomy — the physical part, the landscape — whereas 'mind' is the physiology — the biochemical working part, the ecosystem.

SOUL is our essence. Soul originates when we were one cell that became many cells, and each of the cells, in spite of their specialization, agree that they are part of a team that is 'me'. They each have their own experiences and viewpoints but collectively share.

The intelligence at the cellular level contributes to who and how we are but, for most of us, it is so subtle that our minds cannot perceive or control it. Soul is all-encompassing, the great connector between the parts of who we are internally and between us and the material and spiritual world outside us. It is all that we are and our potential, both good and bad. It includes the other parts — heart, body and mind — for they function as part of our unique creation, made up of molecules from the stars which resonate with a unique energy. It is at this energy level that the soul might see and know the spirit, for the soul is the part of us that communicates with the intangible.

We often see soul written as if synonymous with spirit, which doesn't detract from the main flow of this conversation

— "save our souls" or "your immortal soul", and it can even be equated to 'beings' as in "twenty-nine souls on board".

Spirit, some say, is an entity that abides in us and is created and delivered at some point when we come into being. If we are God-believing, we believe that God gave us that spirit when we were conceived. For those who don't believe in God, there is a wide spectrum of beliefs as to how the spirit world works, including the belief that there is no spirit realm. If it was possible that the spirit arose out of the energy that formed when we were conceived, then the spiritual world would depend on the material world energy for its existence. This wouldn't be consistent with all that we know and believe to be true, that spirituality is not contained nor controlled by the material world. In addition, if the material world could create spirit, then the end of material, at death, would mean that the spirit would die too. For this model to work, we could think of Mind and Heart involving anatomy and physiology, and Soul holding *them* by one hand and Spirit by the other.

So heart, mind and soul are quasi-body parts that help us to visualize what they represent, which are *emotions*, *thinking* and *existential essence*, respectively. The closer you look, the more each definition is similar to the others due to each encompassing all parts of who we are. Each of heart, mind and soul want to be with the others; it is only us who split them up.

When we look at these parts of us diagrammatically, we can see a representation of how the parts relate to the whole. In ancient times, there was apparently more attention paid to the spiritual aspect of life. As we have become more comfortable in our lifestyle and more knowledgeable in our understanding, there has been less interest in God and spirituality in the traditional sense of religion. If this is a trend and it continues, we might see a progression that is illustrated in Figure 21. Some people might respond to this in dismay, but it is only our most recent history, a hundred years or so, that has manifested this trend, barely a geological blink in humankind's story.

Figure 21

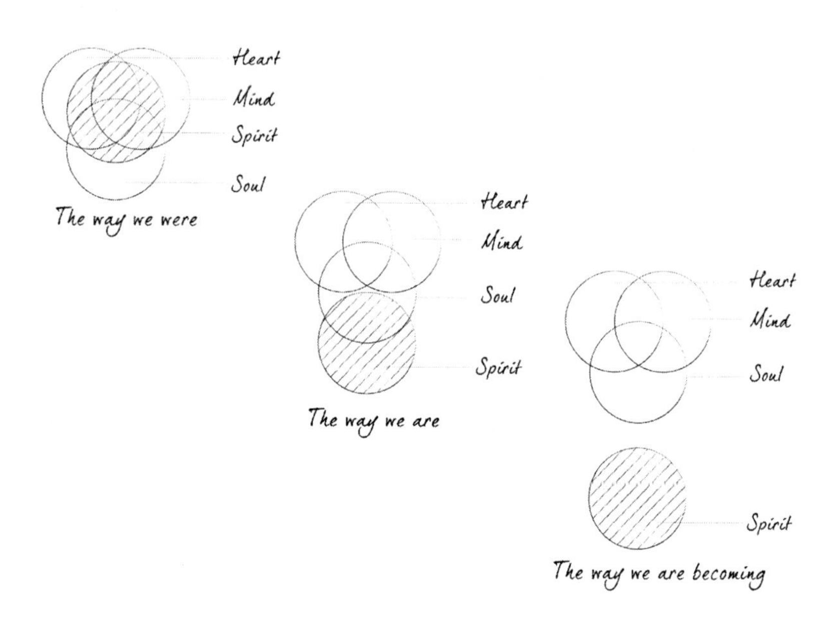

Circle diagram of heart, mind, soul and spirit

7

Computers, Machines and Robotics

Mind Control

Our most primal neurological reflexes are so ingrained that multiple species share the same processes. These processes work by using neurons as the physical, molecular rails for electrical energy to glide down, forming paths in our brains, and new paths can be created and old paths can be disused. Add to this the idea that unseen electromagnetic waves are energy that can have an effect distant to their source — for example, radio waves go from the transmitter to our radios, X-rays go through flesh and make a radiographic image, ultraviolet rays are used by plants to convert CO_2 to carbohydrates and can tan or burn our skin. Combine the two concepts, that brains have electronic pathways and that energy can be wireless, and you have the notion of remotely manipulating electromolecular pathways in the brain, or in essence a way to control how other people think.

If you can control others by transferred thoughts, you then have ultimate control. You wouldn't need an army because you could simply instruct the opposing army to destroy itself.

This idea was poignantly expressed in John Wyndham's book *The Midwich Cuckoos*. The cuckoo bird lays its eggs in another species of bird's nest for those parents to raise the babies, and the theme of the book is that an extraterrestrial species of hominid similarly placed children with extraordinary abilities in a town on Earth. After a human fired a rifle at one of the children, another child sequentially seized control of the minds of those around him until he found the guilty party hiding in the long grass, and instructed him to shoot himself in the head. Threat resolved. Recently, the movie *Push*, written by David Bourla, showed people with exceptional talent who could instruct others to shoot themselves. These examples are one-on-one instructions by spoken or telepathic communication.

The ultimate weapon would be the nuclear bomb of thought transference, where everyone who received the message would do as they were told. As with everything, there would be a light and a dark side of applications. Obviously, manipulation and harming others are part of the dark side. On the light side, you could instruct people to harmonize, cooperate, work for global justice, cure cancer or find the answers to the secrets of the universe. One response to all this would be the outcry about loss of freedom and personal self expression, but it would appear that people with power and money are already affecting how we think by lobbying and advertising.

If benevolent people discovered this process, continued to be peace-loving and didn't abuse it, then it could serve as the universal parent that humanity needs. Inevitably, though, the dark side would discover or steal the power and the battle between groups of people would be taken to new heights. Perhaps evolution has built into itself a regulatory process whereby Sapiens cannot access processes with which we would quickly annihilate ourselves. We learn in small bits, grow with the new knowledge, accommodate, survive. For those who believe in God, we might conclude that God only allows us to know that which we are ready to know without destroying ourselves. And we're always playing to the limit, aren't we?

Evolution of the Assembly of Metal Molecules

Evolution started small, then grew over time on the basis of competition. Computers started huge and are shrinking quickly, the opposite dynamic, but also on the basis of competition. There will come a crossroads where the tiniest computer part will meet its functional counterpart in the biological world.

The miniaturization of computers is unknowingly growing towards the recreation of life, where molecular structures perform tasks. We are unconsciously reproducing ourselves in a form that is faster, more capable, more reliable and more able to complete tasks that we cannot. Ultimately, assuming the new life in robotics will also be given sentience, we will recreate the perception of ourselves and, even as we discover what we are doing, we won't know why.

One possible reason for this trend is that it is part of the progressive march on the road to discovery of the universe, and a part of that exploration involves space and time travel, which are probably not suitable for biological beings but are well-suited for robotic ones.

Computers are basically very simple things — they decide 'yes' or 'no' by way of a '0' or a '1', but they can make this simple decision a billion times a second. So, rather than think like we do and come up with an answer, they plough their way through every possibility, one at a time, to come up with a solution. This is good for mathematics, accounting and storing enormous files, but doesn't come near rational thought.

As we continue to miniaturize electronics and computers, at some point we will have to bridge the gap between the worlds of zeroes-and-ones and thoughts in order to produce a thinking machine. 'Fuzzy logic' is the realm between 0 and 1, where we do most of our thinking, and computers will need to learn that process before they can think like us.

We are often in the struggle of turning infinite shades of grey into black or white, a zero or a one, but that effort is itself in the grey zone, an area traditionally outside the realm of computers. When we ask computers to give up their strength, black or white thinking, to enter the grey zone and swim in all the possibilities, one wonders if their speed and clarity will be lost. They could become like us in ways we hadn't anticipated.

Computerized Models of People

Similar to the millions of decisions that a computer makes to reach the endpoint of an answer of some kind, there are millions of decisions that the genes in a human body make to reach the endpoint of stable life. Once we know the products and effects of the entire human genome, we will be able to construct a human model by computer, putting in the personal variables for any one individual. Then it will be possible to play with the computer model of the human body by having the model 'eat' an apple, or rather the digitalized chemicals that make up an apple, and see how those molecules are absorbed, metabolized and excreted at the biochemical level. Or we could have one cell go crazy with runaway replication — also known as cancer — and see its behaviour, then the effects of anti-cancer drugs. In a computer model, it'd all be theory, of course, but still, the more accurate the input data, the more informative the conclusions. This would drastically decrease the need for animal experimentation, complete with its limitations in extrapolation to humans, and would aid in evaluating how toxic a drug is. Also, the model could be aged to see the long-term effect of a drug or disease condition, which is really an unknown parameter for new drugs.

If every single variable that is possible within a human could be categorized and quantified by a series of 0s or 1s and these were assigned an electronic format, like matrices in three or more dimensions, it would then be possible to describe a person by a mathematical curve. Gene expression through protein mediators, current body anatomical configurations, limits of physiological parameters and every measurable aspect of a living body would be inputted. Free will, unpredictable choices and random real world occurrences would still make each person and their life unique, but the potential of their being could be estimated. This would give people a realistic idea of where their talents, strengths and abilities are, and act as a counter-evolutionary force, allowing them to find their optimum place in society without the survival-of-the-fittest competition. Beyond that, competition could still exist but grow into a neuroplastic maturity, a stretching of the mind that helps with social order without creating second-class citizens.

Parallel Evolutions

Computer technology has involuntarily and subconsciously mimicked cerebral evolution because that is how we think, so we plan and make computers to think like us. Now the process is coming full circle, as we can look at computers and start to understand how our brains work.

All thought relies on biomolecular form, to transfer energy, probably in wave form, along or through molecules. Thought might be purely electrical or energy, but memory is probably solid (or energy hovering around solid molecules). Just as computers can hold a thought as a memory in electromagnetic form, we likely can hold a memory in electromolecular form. We have pipeline caches where we can find the recent memory quickly. This has been an evolutionary advantage, whereby the neurological paths in our brains could be accessed quickly for hunting techniques, survival and reproduction. We still use them today but often they stand against us, where a solid, molecular path in our brains relays a message that is negative to our well-being and creative expression. While we cannot erase the old path, we can make new ones, though it takes great discipline. The inclination is to slide back to the old familiar way. We also have RAM, random access memory, where we can go directly to the thought we need, and sequential memory, where we bulldoze through all the thoughts from stimulus to conclusion. What we don't know yet is how the neurological nudging against the edge of the molecular thought can recreate the memory into our consciousness again. To be truthful with ourselves, we don't know what consciousness is, nor thought, nor memory, let alone how they work.

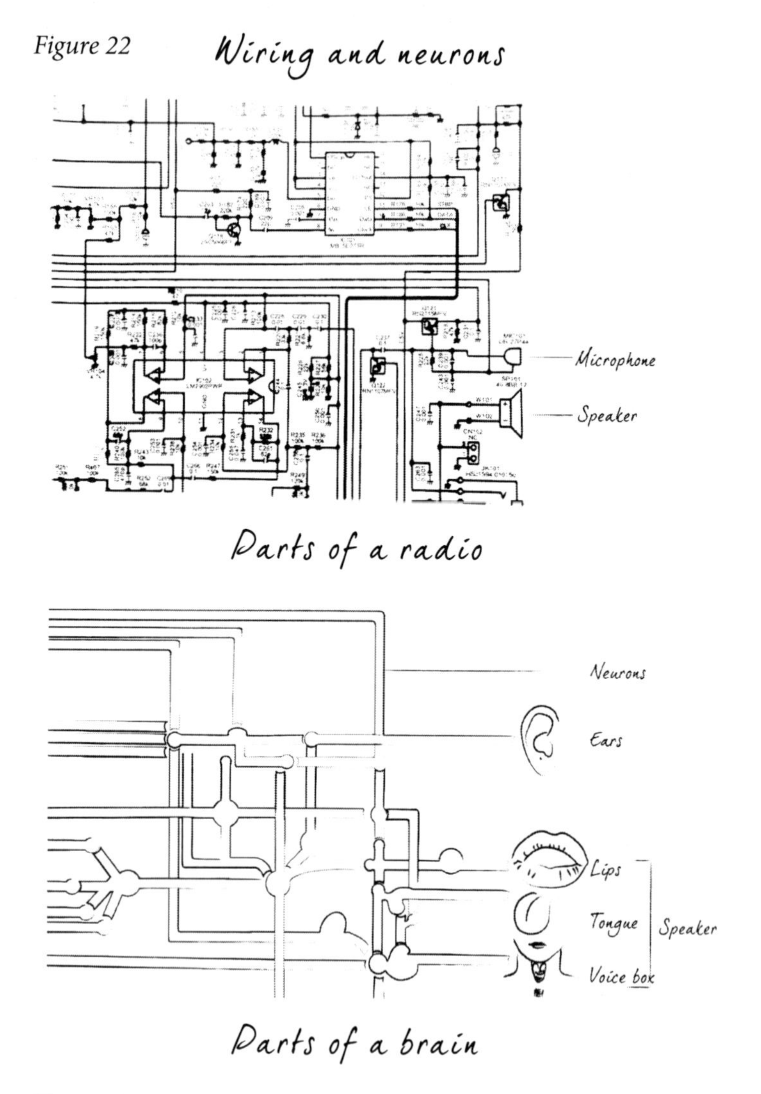

Figure 22 Wiring and neurons

Fine Tuning?

When we think, we access several compartments in our brain at one time and network the various components to produce conscious thoughts. If all the compartments were allowed access to each other all at one time, it'd be cerebral chaos, a cacophony of unrelated memories and processing — it would be insanity. So in order for computers to make the leap from their fast 'yes' and 'no' system to one of reasoning, there will have to be a mesh, a network or a three-dimensional array of circuits that interconnect, along with intersections that decide where to send the electronic energy. Maybe the working concept of miniaturized silicon chips at all the crossroads within a graphene[8] mesh would mimic the brain.

A four-legged robot called Big Dog, made by Boston Dynamics, has gyros and computers with sensory and motor ability at each leg, all connected to a master computer, a CPU in the body. Its purpose is to carry a load over rough ground. During a test, it was put on slippery ice. As it slid around, and each leg quickly thrashed to regain balance, it looked like a long-legged beetle imitating Bambi on ice. It successfully regained its balance and continued walking. When it was struggling and nearly falling, it was easy to have emotions about it, wanting it to succeed and to not be damaged. This would be transferred emotions that we would have concerning a deer in the same situation — it was a sort of compassion. It would amount to compassion for a bucket of bolts and wires. This actually makes sense, because the closer we can mimic life, the more we will embrace that imitation into our sphere of experience. We already have emotions about our cars, our houses and our golf clubs, so when a robot acts alive, it seems natural that we will have emotions around it.

Alive but not Spiritual Robots

Now if we consider that an emotion is preceded by a thought, and that thoughts are the result of multiple neuron interactions — akin to computer chip — it could be possible

8. Graphene is a single layer of carbon atoms laid out in a uniform pattern that looks like a chain link fence. It can transport electrons; its unique features make it a candidate for quantum computers.

that robots will eventually have emotions. All that is necessary is to program into the robot the neurological information within the amygdale and limbic system of the human brain.

If you can imagine all this taking place in the foreseeable future, then you have to wonder if thinking, feeling machines would be able to contain a soul. Every person has a soul, the interface between the molecular you and the non-material you. This is contained in space for the time that it resides in your body. The spiritual world and God are beyond the confines of our spacetime universe, which makes sense from the definition of God, without limit, and from whom we get space and time. Perhaps the nonGod people can consider this a temporary theoretical definition to allow for this conversation to flow.

At any rate, everyone has a soul — just because you can't confirm it scientifically doesn't mean that it doesn't exist. We can't prove the existence of a graviton, but gravity still exists.

Considering spirits, in the God camp we have it easy — we say that God gives out the spirits. He's the creator of everything — quarks, mesons, planets, souls, me and you. He creates the spirit and it gets to live in a body. According to biblical accounts, God knew you before you were born, which suggests that you were a spiritual being before becoming a material one. It also says that He knows every hair on your head, indicating that the creation of your material being was very exact and unique, which agrees with our genetic and biological observations. A spirit is immortal going forward in time, but how far in advance was it created before being given a body to live in — one second, one year, one million years, or some 'distance' apart from time? We perceive that eternity goes both ways, into the past and into the future, but immortality also bypasses time and so might go both ways too. In any case, there are reasonable indications that it's possible that you had a spiritual form before your father's sperm cell met your mother's egg cell to start the material you.

This spirit is perceived by the soul aspect of our being, but a soul — by the definitions within this conversation — can't turn it around and create a spirit. So we could conclude that something made by people that looked, acted and felt like a person could mimic a soul, but wouldn't necessarily have a spirit. As everything can come in gradations, as a virus acts like life but isn't commonly accepted as living, so perhaps an advanced robot would challenge our understanding of life. A hybrid man-machine or a sentient robot might make us wonder if it could harbour a soul.

The immediate reaction of Sapiens to this idea, regardless of any religion, wouldn't be spiritual but simply species arrogance — nobody can be as good as us. The religious would quickly point out that robots are not made by God but manmade, which would be true, at least in the beginning. Eventually, though, robots could make themselves without our help, which still strictly wouldn't qualify as life, due to the missing 'reproduction and growth' aspect and lacking meiosis and mitosis, but would be self-perpetuating.

The argument would fester around the comparison of two beings made of a sustainable collection of molecules, one a biological lifeform and the other a technological lifeform, with both performing many of the same functions, possibly including sentience. Perhaps our destiny is an extension of the heterogenic mixing of races and cultures that produced progress in civilization, whereby a mix of Sapiens and robotics will cooperate to harvest the wealth of energy of the stars.

Some in the nonGod camp might think that all that an individual is — with the matter and energy of one hundred trillion cells, the totality of those atoms and molecules and cells, that living being — contains an intelligence that generates a spirit from within. This puts material limitations on a limitless spiritual possibility. It also begs the questions, how do you care for a spirit if it simply arises from cells and where does the spirit go when the body dies? If the spirit dies too, then there couldn't be any permanent union with God and no relationship with infinity and eternity, those things that we yearn for, and probably the whole reason that humans started their spiritual inquiry, many thousands of years ago.

Robots Will Create Social Problems

We are familiar with the social problems of sexism and racism, and there has been some progress in equality. Still, we haven't really addressed speciesism, where Sapiens — by virtue of arrogance or ignorance or interpretations of words from holy books — declares, without room for discussion, that we can use and kill whatever we like, but not each other. We do kill each other, but we tend to set clearer limits lest we end up on the wrong end of that dynamic.

There is an opening scene in the movie *The Last of the Mohicans* where a man and his two grown sons hunt a stag, kill it, then pray over it, honouring its existence, wishing its soul safe journey and thanking it for its body that will yield meat, clothes and tools. This type of scene was revisited in the movie *Avatar*, where a Sapiens in an alien's body learned the traditional ways.

Today, someone else raises the pig or cow, another slaughters it, another butchers it into little cuts of meat, wraps it and puts it into the display cooler where it becomes an object to eat. Same taste, more efficient assembly-line system, cheaper and cleaner, but with a loss of connection at some level. The molecules would all be the same.

Do molecules change when spirituality is added? We are eventually what we eat, molecule for molecule, so you have to wonder: if molecules that are acknowledged as life are then consumed, would they then be building blocks in a body that somehow resonate differently?

The study done on water by Masaru Emoto, mentioned earlier, showed that, when subjected to different stimuli then frozen, water's crystal structure was different depending on whether the stimulus was negative or positive. So it is possible that molecules arrange themselves differently or manifest their energy in different ways, depending on how they are treated.

The point is that we can still eat meat and be part of the big ecosystem, but *how* we do so impacts who we are, literally.

Social Issues with Robots

Beyond all this, if sentience is the ability to have feelings and emotions, and we reflect on the previous discussion about robots becoming sentient beings with thought and emotions, it brings into question whether such beings will insist on a say in their care and in their destinies. As is our usual pattern, we will resist any change to the status quo, where human is king and machine is invention, and insist that machine just keep quiet. Both humans and robots would be made of the same stardust, of molecules that reverberate together, both would possess organ systems that cooperate for the functioning of the whole, and both would think and feel. Yet the humans will, in the beginning at least, claim the sovereign right to be master.

The God camp would argue that God, who is perfect, made Man in his image (spiritually at least) and gave him all the beasts of the earth to do with as he pleases, and that robot is made by man and so there is no comparison.

We would have to acknowledge that humans are alive, can biologically reproduce and have a spirit, and that our concept of robot at this time does not include those variables. More importantly, it is not so much whether another being has feelings but our response to their existence that demarcates our humanness.

Our response to the universe's challenges is as important as our place in it, and our usual response is to be slow to change. We used to believe, even just a hundred years ago, that animals had no feelings. Women in the Canadian province of Newfoundland and Labrador only got the right to vote in 1949. Women in England in the not-so-distant past could be beaten by their husbands with a stick no thicker than his thumb (thus the 'rule of thumb'). Martin Luther King was murdered in 1968 for seeking racial equality. These examples of speciesism, sexism and racism show that these dynamics are all fairly current, so we're nowhere near ready to address the impending robotism.

Perhaps a dictionary in the future will say, "Robotism: where collections of self-sustaining molecules that are not biologically brought into being are not honoured and respected as sentient beings". Of course, this is not relevant yet, because computers haven't yet been programmed with emotions and cannot yet think.

As you work your way through the macabre concept of robots being so molecularly complex to be able to harbour a soul, it opens up many other difficult questions. If collections of molecules can have a soul, then what about all the other animals on Earth? What about plants? Different spiritual organizations have differing viewpoints about who or what has a soul.

If sentient robots were successful survivors on Earth, more so than hominids, they could become the ideal intergalactic explorers, but might not have any use for the lifeforms on Earth. We see the beauty and the aesthetics of all the creation on Earth that we are destroying, whereas robots would leave Nature alone unless it got in the way and then probably bulldoze it away without compassion — *unless* through their higher intelligence, they knew that the atmosphere and the global climate are determined by the total biomass, which in turn could affect the stability of the metals that they are made of.

The key survival concerns for robots would be access to minerals to construct things and sources of energy. Also, if the evolutionary transition is from geo to bio to techno, and this proves to be the ultimate survivor, then it could be that this has already happened out in the universe. This might give us an element of caution in attempting to contact aliens, as Professor Stephen Hawking has warned us.

Robots into Space

There have undoubtedly already been a million varied attempts before molecular organization could finally result in the successful creation of the building block of living beings: the biological cell. One cell likely started all of life on Earth and one cell started all that is you. It repairs itself. It has all the intelligence necessary for life in its DNA, the mastermind of its existence. It controls what goes into and what leaves the cell.

This is where electronic miniaturization seems to be heading — to mimic life but without life's vulnerabilities. In essence, we have already been using the strategy of machines to do our work — our cars move us, our telephones move our voices, transportation devices like freighters and trucks move our food and drink from faraway places to our tables, and our computers think out problems and memorize things for us. We are already sending robotics into space like orbiting satellites and the Mars Exploration Rovers, and this will be extended in time, in space and possibly in dimensions, with progressively more sensory capabilities. The idea is to send extensions of our biological processes without risking our fragile bodies to the known and unknown hazards of space and time. With miniaturization, we can send beer-can-sized machines into space at a fraction of the cost in rocket fuel, or at the same cost in fuel to send a full-sized rocket, and sending 10,000 at a time. This reminds us of an oyster laying 100,000 eggs, increasing the odds of successful survival. This is thinking egocentrically, with us at the centre.

We could think the other way around — that we are not the inventors of this idea of sending seeds into the galaxy and that what we're doing is how we got here in the first place. DNA is an impossibly unlikely spontaneous creation unto itself and might have been planted here. However, this only defers the big question of how it arose, by transplanting the question to another place in the universe where the same question can be asked again. How did life begin *there*?

Travel

We are immersed in a car culture the world over. Everyone wants one, along with the power, comfort, safety, prestige and independence that they bring. We think of the modern car as a machine, but it could be considered a partial robotic device which still needs full-time supervision with many of its functions occurring automatically.

Internal microclimate control and music-playing abilities have been with us for decades, but are getting fancier. Throttle control is now by electronic fuel injection instead of simple carburetion, so the changing mixture of air and gas at different altitudes is automatic. Cruise control is nothing new, incorporating a degree of artificial intelligence (AI) to operate acceleration, with directional or braking control developing but not commonly available yet. Some vehicles stop the functioning of some of the engine's cylinders when their power isn't needed, thus saving on fuel, and others change the traction from two-wheel to all-wheel when sensors detect that the extra grip is needed. There is now a preferential braking system available that allows certain wheels to brake harder than others in different circumstances — for example, when braking hard in a straight line forward, the back brakes apply more firmly, to negate the nose dive on the front end. There are experiments taking place with the car being programmed for a destination and then not needing any human input along the way — total automaticity. GPS already knows the speed limits of the road you're travelling and emits a warning when that speed is exceeded, so this would be one way to set the basic pace of a vehicle.

There would have to be other variables added to that, such as sensory awareness of objects around the car, then computer feedback to the acceleration, braking and directional controls. An example of this already exists with partially automatic parallel and reverse parking assist systems ('partial' because the driver still needs to verify with the computer and keep a foot on the brake).

Another model for automaticity is to have all the cars communally owned and on rails, so you could stop one, like a taxi, get in and program the vehicle for its destination.

Having them on rails would decrease the need for directional control, and the acceleration and braking would work like some public transit systems do, with no human conductors. Initial infrastructure would be expensive, but the monetary savings through accident avoidance, the increased efficiency of computer piloting and the decreased dependence on oil (if the system is electric) would offset that expense.

The point of this conversation is that personal vehicles run by AI would make them *fully* robotic and this seems to be the path we're on already.

As Katie Alvord notes in her book *Divorce Your Car!*, our relationship with the car has evolved from a beginning where the first people to attempt to drive a motorized vehicle were arrested — they were piloting noisy, smoky, scary metal beasts. Then, as cars became privately owned and less experimental, they were often greeted by hurled stones, as it would be the wealthier who owned the cars and the car was seen as an obnoxious, forceful extension of that wealth. Mass production lowered the cost of owning a car, they became commonplace and accepted, then squeezed out the horse and the rest is history — a very short history. Now we're on the verge of robotic personal vehicles.

When my daughters were small and we had seen *The Wizard of Oz*, I told them that the lions, tigers and bears of our world were cars, trucks and SUVs. Sure, you might worry about a pedophile or a real cougar (where we live), but on a day-to-day basis, it is a motor vehicle that is most likely to kill or maim you, either from striking you as you walk down the street or by hurting you while you travel in one.

It's not currently possible, but the *Star Trek* model of 'transporting' people from one place to another would revolutionize our thoughts on travel. As long as we cherish the paradigm of sitting in a metal box on wheels that is propelled by small explosions in an engine, we are slowing the expansion and exploration of alternatives.

It is difficult to see how a *Star Trek* transporter can disassemble and reassemble a body without a little death in-between, but maybe it represents a concept only possible in another dimension, or a break in the space-time continuum, one that exists here and now that we haven't discovered yet. We will likely find that everything that we ever wondered about has always been with us.

Whatever we can conceive can come into existence.

The *Star Wars* model of holographic communications could be possible. It would be a virtual construction of the senses, projected into a hologram, and the hologram could be sent anywhere light goes. The trick would be to have ongoing perception and sensory information-gathering to feed back to the person of the hologram, so that you could stay here but be there, in a sense, which would be virtually the same thing, no pun intended.

The movie *The Matrix* showed this in yet another way, where people were plugged into a shared virtual reality. There are video games online where people all over the world can play a shared program. When you add the ability for you to make the others feel and also for you to feel them, then you are there — full sensory, virtual reality — which is in phases of creation and experimentation now.

You can imagine someone sitting at the computer with probes coupled to body parts, some to give information like temperature, heart rate, perspiration, touch and pain, and some to receive that information from others. The virtual reality then becomes a step closer to reality. You could be anywhere there is another electronic terminal, without ever leaving home.

Travel and transport is essentially an energy issue about how to get a person or material goods from A to B. Energy is usually paid for with money, which is usually gained by energy expenditure. The machine necessary to transport the person or stuff takes energy to assemble and create and then energy is needed to make it work. It's a cycle and, like the money one where 'you need money to make money', so it is that you need energy to make energy — at least in our current paradigm.

Figure 23

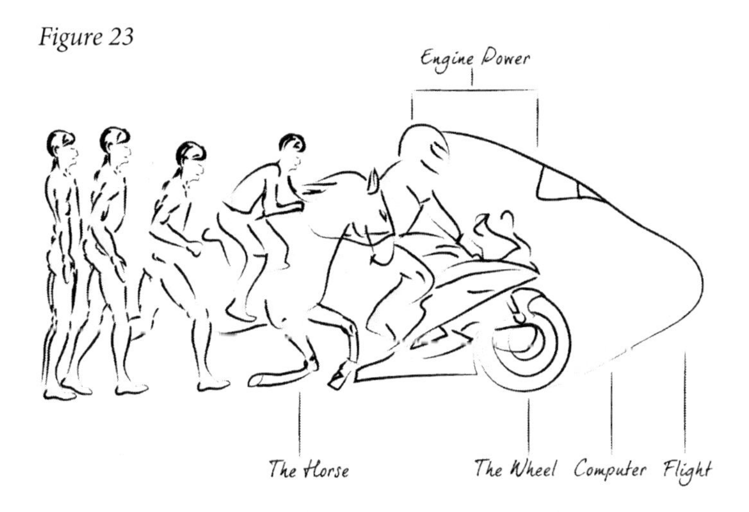

The evolution of speed –
going from here to there, faster, is one way

Perhaps we need to shift our concepts to be like those of a sailor who harnesses the wind, and harness the energy around us. The sunlight, the wind and the seas' tides and waves contain more energy than any amount of burnt earthly carbon could ever hope to achieve. It might come to pass that we let those powers turn on themselves to create an energy product for us to use — for example, where the sun's energy is used to create the wind that drives the waves that propels the generator of electricity.

Soviet astronomer Nikolai Kardashev found himself thinking about such ideas and developed a scale addressing the levels of civilization, based on their mastery of energy.

These Levels of Civilization are:

Type I
harnesses the energy of the planet, controlling sunlight, volcanoes, earthquakes.

Type II
controls the energy of the sun, making them 10 billion times more powerful than the Type Is.

Type III
controls the energy of the galaxy, with 10 billion times more power than Type IIs.

Physicist Michio Kaku says we may attain Type I status in one to two hundred years. Thus, according to him, "our own civilization qualifies as a Type 0 civilization (i.e. we use dead plants, oil and coal, to fuel our machines)."

Figure 24

We are cavemen in cars

8

Spirituality

SPIRITUALITY EXISTS for many people and so it is part of our reality and, if that is true and relevant, then it makes sense to include spiritual aspects of life when attempting to unravel the mysteries of the universe. The intent is not to preach on religion or to convert people to a specific religion, but to include spirituality in the discussion, in order to keep the search for answers as broad as possible.

The Big Event

The coming into being of consciousness was an event that allowed hominids to become human, and was the gateway through which all humankind's history has progressed.

The idea that 'something happened' to us at a turning point many years ago has been presented by Arthur C. Clarke in his book *2001: A Space Odyssey*, where two groups of hominids meet every day with water between them and bellow threats to each other, defending territory or some primitive motivation. Then one day, this enormous black rectangular monolith appears. We intuitively surmise that someone made it and someone put it there. Sometime after its appearance, the hominids on the side of the water exposed to the monolith cross the river with sticks and clubs, and the leader of the other group is killed. Something big happened in the lives of these beings and there was no looking back. We conclude that the presence of the monolith had caused this change, and represents the notion that there has been a trigger for us to step out of the animal world into one of consciousness.

Another viewpoint of this event is the story in the Bible in Genesis, where "God created Man".

Figure 25

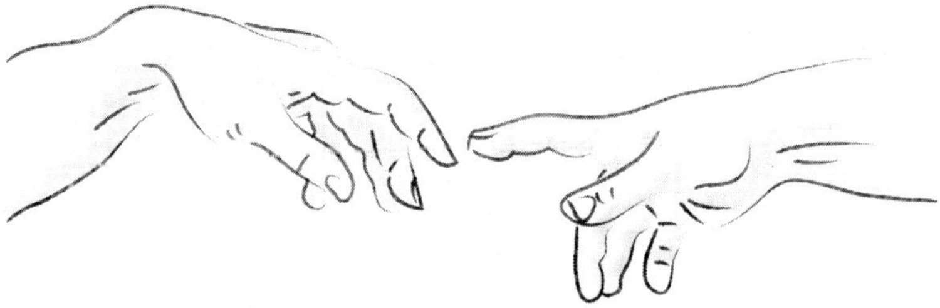

Michelangelo's The Creation of Adam, a painting of God and Adam's hands nearly touching, is on the Sistine Chapel ceiling

It's a poignant allegory to the concept that God has always existed, but humans became aware of Him at some finite point in time. The trigger could have been akin to a shiny black monolith, which points to beings from another planet stimulating our development and just moves the question from this planet to another planet. If it had been spontaneous, most human brains would have to have had the inherent ability to grasp the concept once it was presented to the others. So it wasn't an organic evolutionary thing at that time, but the understanding of a concept, using the cerebral machinery that was already in place. It is probable that the brain had been physically growing over millennia, in response to better nutrition and the demands placed on the brain by competition. Once conceived, the area of the brain necessary to harbour the concepts was exercised and grew in ability. This is rather like the concept of a man running the unattainable four-minute mile — once Roger Bannister broke that barrier, everyone was doing it.

Evil

Evil could be a force unto itself and not need anything else to compare it to, but if you go back in time before there was life of any kind, when the universe was composed of molecules that only aggregated into celestial bodies and not into living ones, what would be the relevance of the existence of evil? And when life started on Earth, and there were uncountable numbers of microbes, what was evil doing then? One species of microbe could deign to take over the world, killing other microbes, and this would just be Nature and not particularly evil.

It is only when sentient beings enter the scene that we place a judgment call on actions that hurt living creatures. Just as the existence of light shows us what the darkness is, perhaps so it is that life shows us what evil is. If it is true that evil exists in the spiritual realm and that the forces of good and evil battle it out, then they are still battling today, but with the inclusion of human spirits. Evil can then exist in the scientific, material realm *as well as* in the spiritual, ethereal realm. It's a bit ironic that we have a difficult time including God in both realms but that we can acknowledge evil in both.

So the definition of evil starts with *the threat against our existence.* It can be both tangible in the moment and spiritually eternal, outside of the limitations of time. The vehicle of evil is fear, which found its beginnings in the animal state through the adrenalin responses, and was given a promotion into more complex permutations when higher consciousness came about.

If an advanced alien culture came to Earth and annihilated all Sapiens in a moment, where there was no fear, pain or suffering, would that be evil or just an augmented form of Nature and evolution, but with compassion? It could be like the microbes' view of Sapiens when we rid the Earth of smallpox virus.

It would seem that evil has to include *the stimulus of fear*, where the fearful being receives messages of loss, impending pain, helplessness, abuse and insignificance. In the predator-prey relationship amongst animals, the absence of higher consciousness in the animal state usually precludes the presence of evil. Life there is at the level of hormonal, innate responses without interpretation. So, one more requirement for evil is that it has to include *the existence of consciousness*.

Interspecies Rape

Some male sea otters have been seen to sexually assault young seals, killing them with their penises, and continuing with forced copulation for days after their victims were dead. The seal pups are painfully mauled, to die of trauma and infection, with no regret or change of behaviour from the otters, mimicking evil human behaviour.

Although our feelings about this arise from anthropomorphization and not impartial observations of animal behaviours that are part of evolution, it's difficult not to feel revulsion and a wish to restore some kind of justice.

Interestingly, these otters are the cute ones that float on their backs while using a rock as a tool to crack open a shell. This describes a mammalian with higher brain activity that exhibits ingenuity with a tool as well as a dreadfully peculiar use of force to get its needs met, which tempts us to conclude that *more brain power* is linked to more complex activities, both positive and negative. It parallels humankind's history where the advent of consciousness came *with* art, gods and war.

Without getting into human examples of evil, the feelings we have after reading of the otter-seal relationship shows us that evil has to include *premeditation or choice* by the higher mind.

It's sociologically interesting that fear can arise out of loneliness and fatigue, showing us that consciousness becomes vulnerable when isolated and split off from community or when its defences are worn down. Also, as we frequently see in the news, when a higher brain is not working quite right, as in the myriad forms of insanity, it becomes a breeding ground for evil. Of course, insanity or bizarre behaviour doesn't necessarily harbour evil — it's just easy for us to stereotype those ideas together.

The driving force of evil might not necessarily be an entity unto itself, but part of a continuum of existence. An example would be the creation of life. When molecules first started conglomerating into specific conformations, or structural arrangements, in repeatable patterns that became cells, the first primitive bit of life on Earth, there was a driving force, an attraction of atoms. To make the process repeatable, the cells had to exclude some molecules and reactions in order to maintain the new internal harmony.

The dichotomy between the excluded and the included created the first level of competition which became biological Nature, and is the basis for everything that exists about life. Cells competed against each other for space and nutrients, then multicellular organisms did the same, then populations of living beings repeated the process, all the way to Sapiens, who compete for nutrients, space, power, finances, military power, good looks, mates, you name it — more, bigger and greater, with concern about 'self' being much as it is for other species.

> The natural process of life is to compete and build, with competing forces affecting a balance. If the opposing forces do not create a balance, then growth occurs until a limit has passed and collapse ensues.

This competitive process is a constant force, like gravity. Does that make competition and gravity evil? Not really, because without the former there'd be no life and without the latter there'd be no universe.

On the other hand, it could be that along with other existential issues, evil can take on a life of its own. Similar to the old philosophical challenge, "If a tree falls in the forest and no one hears it, does it make a sound?", we could say, "If evil had no life through which to express itself, does it exist?"

This would simplify the question to whether evil exists as a spiritual entity on its own and found its home in consciousness, *or* to whether there was no evil until there was consciousness. In this case, the existence of the mind allowed for the creation of evil.

In summary, we are looking at:

- Evil existing on its own as a spiritual identity, possibly finding a path to express itself through consciousness, *or*
- Evil being an offshoot of evolution, whereby it was necessary to have:
  1. consciousness
  2. fear
  3. choice
  4. harm

When we think back to our ancestors, *Australopithecus* probably wasn't as smart as my dog, but *Homo erectus* could take a stone and form it into a spearhead, showing the benefits of opposing digits but also of higher consciousness. It took conceptualization, planning, development of technique and learning from experience — a series of intangibles that we have yet to see clearly in other species. They were around for over a million years. At their level of awareness of the universe, of time and mortality, you'd have to wonder if their actions at times considered evil, or were they still innocent of those types of actions because they were still animals acting reflexively through genetic pre-programming?

Neanderthal was aware enough to harbour evil, but did he? Were there members of that species that manifested evil through cold-hearted, merciless, vicious, selfish harm and destruction to others of their kind or other living beings when they knew they had a choice? We don't know, but it was likely, as the dark side of consciousness is the ability to abuse it in the form of premeditated negativity onto others. If so, this gets us Sapiens off the hook, and absolves us of being the first species on Earth to harbour evil. Otherwise, if Neanderthal was fair about invoking justice and equality amongst their kind, or was still too animal-brained and lacking consciousness, and Sapiens was the first to harbour evil, then our simplistic conclusion could be that God created Sapiens and Sapiens created evil.

This concept creates all kinds of theological and philosophical arguments, and leaves our species dethroned from a holy pedestal. The nonGod camp might argue that evil is just a manifestation of human behaviour, which is an end product of the competitive forces that started with molecular attraction, through unicellular to multicellular to organismic to population competition. There will be many viewpoints in-between, with or without God, and with or without the acknowledgement of the existence of evil.

When we are living, evil is a threat against our material being, our personal collection of matter and energy that is a tangible resource to fight over. Was evil relevant to me before I was born and after I die? I would hope for and I'm counting on heaven *not* being a spiritual battleground, where the intangible resource to fight over is spirits. It would seem that life, somehow, is the forum in which both battles occur, for the tangible and the intangible, the former for survival in the world and the latter for survival of spirit.

When we step back and consider these other possibilities, it shows us different paradigms. This could shift our viewpoint from our whole existence resting within the material universe and day-to-day life, with belief in the spiritual realm being optional, into the viewpoint that the spiritual realm needed a tangible place to battle out the struggle between good and evil.

> "We are not material beings on a spiritual journey; we are spiritual beings who need an earthly journey to become fully spiritual."
>
> — John Bradshaw, *Healing the Shame that Binds You*

Without molecules assembled into life, there would be no material trophies for good and evil to keep score. In the spiritual realm, where there is no energy, matter or time, it is difficult to visualize or understand pure spirits harbouring or siding with good or evil. You would think that any spirit, unencumbered by form, would be free to join God (or for the nonGod people, that perfect spiritual place/being/concept), with limitless power, fulfillment, understanding and purpose. It could be that the spiritual dichotomy is God or not, or God versus Self, rather than good against evil. Which brings us back to consciousness as a place where evil came into the world — consciousness the intangible, arising from tangible molecules, the resting place of evil.

Just as we've seen the internal 'push' of Nature within our molecular and cellular structure that can put survival on automatic and can become greed and other nasty traits, so it is with evil — we can have evil in us and be part of its process. Just as the healing process (growth at the cellular level) gone awry can result in cancer, so it is that consciousness gone awry can result in evil — both are potentially within all of us.

One of the processes by which evil comes into being is through lack of self-control — the higher cerebral centres have to manage the negative thoughts and prevent acting them out.

Thoughts themselves can be the breeding ground for evil, and the more we think along those lines the more likely we are to act them out. As the saying goes, "Practice makes perfect." An extension of this is that the company we keep can set the standard for how we think, and create an environment that fosters evil, as illustrated by another saying, "Birds of a feather flock together."

When villages of humanity were one to two hundred people (which seems to be some kind of workable, natural limit, throughout history and worldwide) and there was one person who manifested evil behaviour, that would be a 0.5 percent incidence (1 out of 200). If you then applied that incidence to a city of five million, that amounts to 25,000 people. That would be all the people who *acted out* their evil, and we don't see that many rapists, serial killers and nasty neighbours in each city, so at face value, it is a credit to our species that we're cooperating quite well.

However, if we look at the 0.5 percent evil being spread out within all of us, that would suggest that the mass of evil (if there is such a thing) is growing with the increasing population of Sapiens. There have been many times in our recent history where evil behaviour has arisen in large groups of people, especially in the chaos created by war, which presents a forum that releases evil from its usual containment.

There is the philosophical viewpoint that the strongest person is the one who has the least needs. We always think of the strongest as the one who manifests the most power to get their desires fulfilled and so the billionaire is the representative of that. If we were to compare a street person, probably with some mental difficulties, to a billionaire, it's likely that the latter is the stronger person. But if we include a sage, mystic or prophet, the scale could swing the other way.

How much power did Gandhi generate in the process of owning nothing, and how has his power been manifest in the world since his existence? What about Kahlil Gibran, Isaiah the prophet, the Buddha, the prophet Mohammed and Jesus?[9]

At this point we could summarize that evil can be present in anyone and that personal strength is not the same as spiritual strength. Also that, although evil arises from within people individually, it can also conglomerate into a group of people. An interesting consideration within this conversation is that knowing God does not guarantee that evil won't arise in you — as long as you have consciousness and molecules, you are vulnerable. One would hope that following God's principles for your life would deter evil, but every day is a new day and you can't sail a boat with yesterday's wind.

From the Beginning until Now

Let's take stock of what we know. Atoms came into existence. We don't really know how, but they're here, so we can assume there had to be a force that pushed them into existence. All of the energies of the universe came with matter and, likewise, we don't know how. Quantum forces, gravity, heat, light and magnetism are the players in our physical world and we have some observations about them, we've seen their 'shadows', their responses and effects, but there is still significant mystery.

From the mixture of the atoms and the energies arose a cell, a reproducible biochemical microbe that was the first representative of life. We don't know what pushed chemicals into life. Even if you subscribe to randomness as the process, that doesn't answer how randomness came into being, why it was used as a way to create life, and, still, why life came into being. After a few billion years of development, plants, then animals, then 'higher' animals, then *Homo sapiens* came into being. We can see the pattern repeated over and over, that through competition and evolution, more species came to exist

9. Some Christians will be indignant seeing their god, Jesus, grouped with these notable people, but we're trying to keep the conversation flowing here, and I'm sure that He won't mind. God is not threatened by people or their thoughts.

until we did. We have consciousness and thoughts and we don't know how these exist or how they work. Within the framework of all these marvellous riddles there is the question of 'why' that science does not seem to have an answer for. Simply, science looks after 'where', 'when', 'what' and 'how', whereas 'why' is more of an existential question that searches for reasons.

Imagine there is a 'force' that is behind all of the mysteries listed above and that you are part of it, and so is everyone else. If you are a God-believer, it is easy to let God be that force. For the nonGod camp, you could imagine that behind all these forces of the universe there is a benevolence, a positive purposefulness that includes intangibles like love, acceptance and completion, and that you are part of that. After all, you are Sapiens, which, as far as we know,[10] is the most spectacular collection of quarks in the history of the universe. Forget the argument of design or randomness for a moment — you exist which is amazing and so has to be good. Now if you align your mind, molecules, energies, intents and willpower with God or this All-Who-Is then you turn to where you arose and complete the circle. Jesus said, "Love the Lord your God with all your heart and with all your soul and with all your mind. This is the first and greatest commandment." He was quoting from 3,500-year-old writings, from Deuteronomy in the Old Testament, the Judaic Torah. The ancients, including Jesus, knew the essence of all the contents of this book you are reading now and were trying to tell us.

Different Ways of Looking at Things

There is a practical, psychological technique of creating a cerebral environment and then stepping into it, becoming it and claiming it. Examples would be to think like a wealthy person and begin to make decisions that cause your life to improve materially, or thinking like a world class athlete, invoking positive mental images that guide your training and

10. We don't know any aliens.

discipline into superlative performance. So it can be with peace of mind and happiness, that when a person reflects with appreciation and gratitude that he or she has a great life, that their needs are being met and will likely continue to be met, and that they can afford to be generous with their time, energy and resources, it removes a lot of competition with their neighbouring humans. That's the science of it.

For the faithful God-believers, when they are on His path, they put Him first, then others, then themselves, trusting that all of their needs are met not only for the moment but for all eternity. By trusting in Him, they are resting in faith that they are taken care of in this life and the next one. They don't have to be stupid about it and *not* take care of themselves and make good decisions, for they are part of Him and His creation and each person has — as Max Ehrmann notes in his classic poem "Desiderata" — "a right to be here". In this sense, God becomes the limit to the believers' fathomless desires and drives and He can be the missing regulator over Nature's primitive push to hoard and focus on self. NonGod people have regulators too, such as self-discipline, generosity and compassion, but they'd have to believe in a way of being that supports other people first in order to have a path in which to impart those philosophies.

The conversation of 'others first' began as a way for an individual to experience another way of being that benefits both the person who holds that view and the one who might benefit from it. From here we can expand that viewpoint to include many people, entire societies and ultimately, the bulk of the species of Sapiens.

The Throne in Your Head

This is a simple visual representation of the central core of power and decision-making in our lives — a little king or queen's throne, like a small piece of dollhouse furniture, virtually mounted in the centre of our heads. On it sits the commander of all the rest of our thoughts and activities. For most of us it is Me who sits there: I am the most important person in my world, I call the shots, I have needs and wants that are perfectly acceptable and if I don't get them no one is going to give them to me. We could call this Ego, for all its good and bad reputations, and the driving force to keep Ego on the throne could be the same as for quarks coming into existence and for DNA to form — the primal force of nature and evolution. Every individual living being has to take care of itself. In that process, competition ensues and neighbouring living beings can become a help or a hindrance to survival.

As we observe, the evolutionary system has delivered us to this point in time and space, for better or for worse, and those species that are currently surviving have ridden on the backs of their successful ancestors. However, we are now in a position that we have never been in before: overwhelming success. Sapiens has beaten off the Four Horsemen of the Apocalypse, and is essentially without regulation in population growth, resulting in a top-feeding species becoming very top-heavy in the balance of life. We are rapidly reaching a mathematical limit to the number of humans that the planet Earth can support, and unless we change how we think and act, that limit will be exceeded and collapse will ensue.

For humankind to survive as one organism and to move on to achieve the realization of its dreams in unravelling the mysteries of the universe, there has to be a more universal commander on the throne in our heads.

For the God camp this is relatively easily accomplished by putting God on the throne. When you believe that God is the source of everything and our ultimate destiny, it makes sense to believers to put Him first so that everything after that can follow the guide to the best path of life. Or to put it another way, if our matter, energy and spirit are part of God, then we might as well acknowledge our immersion in His domain by actively choosing it, in order to know how best to live. While this might sound good, what we see around us is that people who believe in God are seemingly diminishing in numbers, they've never agreed on who God is, and the ability to see divinity in all things is getting scarcer.

If the incidence of evil in the world is being kept in check or is decreasing relative to the growing population, as mentioned above, and there is truly less divinity in the world, then Sapiens is on a new path. At face value this looks like social order is being attained by secular organization, however if evil has some kind of latent constant (like the 0.5 percent), then a breakdown in policing could release nasty behaviour on a scale that this planet has never seen before.

The nonGod people have no universal substitute to place on the throne of our minds, and they haven't seen or accepted a role model from the God group. The most powerful manmade force on Earth is money. While this guides our daily actions, both as individuals and as nations, placing it on the throne of our minds won't save us. In fact, as the growth of most economies is linked to population growth, money is tightly associated with our impending crisis.

For those who don't believe in God, there is the challenge of having something or someone that you can trust implicitly to sit on that throne and have first say in all aspects of your life. The substitute that we put on the throne could be worse than our egos being there. For the nonGod people, it's difficult to assign someone or something that wouldn't falter when trials and tribulations come. A hero or saint might not be tangible enough; a family member or family itself is in the same boat as we are, so that could be a circuitous worship of Self again; a country is too big and could use you up, tread over you and leave you forgotten; philosophies come and go.

Perhaps a scientific descriptor of God would work, like All-Who-Is, which would mean all the matter, energy and spirituality in our universe. Or perhaps some people would be content with a love of Sapiens, or humankind, to be on the throne in our minds. Or just Love, and as we are part of that, taking care of others would not exclude us from taking care of ourselves. It is the order of care that determines whether competition or harmony ensues.

> For better or for worse, the 'me first' mentality contributes to the forces of evolution.

A more scientific version of the Golden Rule, "Do unto others as you would have them do unto you," is based on extrapolation of our own experiences. We know what we like and don't like to experience and feel, so we could use that as a guideline to treat others. This wouldn't always work as there are a broad range of preferences — rough or gentle, silent or loquacious, cerebral or physical, as a few examples.

> When I was a teen and borrowed my father's car, I reassured him that I'd be good with it and said, "I'll treat it as if it was my own." He answered, "No, treat it as if it was mine."

So perhaps a step beyond extrapolation and the Golden Rule would be to treat others as *they* would like to be treated. When we consider that there are many people in the world who have suffered maltreatment or abuse, often within their own family systems, and that paradoxically they then seek out similar experiences in later life, they might unconsciously urge us to treat them in the same negative way.

The resolution for the God-believers is once again to treat people as God would have you treat them — as unique and divine creations. The nonGod people could formulate a list of techniques on how to treat people based on scientific, clinical, psychological observations, but it wouldn't necessarily come with a motivator to practice them. Logic alone isn't enough.

Once Ego is off the throne, then the merciless aspect of social evolution is blunted, but the main process of physical evolution, through genetic drift, crisis and selection continues on unabated. One has to wonder whether it's possible to have evolution, observe it happening, be part of it, and still have a say in its manifestations. Can evolution exist with purpose and still embrace those who will not progress to be part of the gene pool?

Allowing God or All-Who-Is to be on the throne of our minds gives our species a common guiding force beyond evolution. Ants and bees are good models for cooperation, but they still war with other colonies. This would be analogous to the organization of two separate religions, like Christianity and Islam, where within themselves there is a certain amount of accord and cooperation, but they still war with each other. If God is love and God is everything, then religions have a ways to go yet before they reach where God would like Sapiens to be, which would be with God on the throne of their minds. War is still evolution at work, and still with Ego on the throne. In our higher consciousness, the key force that drives evolution, both social and physical, is fear — fear for survival, of not being right and of loss of control.

In Love

For those people who have experienced falling in love, the process involved removing Ego from the throne in our heads and placing the loved one there. Somehow by giving up on self and promoting non-self, positive feelings are generated. If it is mutual and both parties are in love with each other, then the giving isn't total sacrifice but a larger sharing, a taste of something bigger than us. What usually happens is that after a year or two, Ego sneaks back up on the throne, lines of defence are drawn, and one self argues against the other for a secure place in the relationship.

If the self is pretty healthy then the sorting out of different perspectives can be attained and the relationship survives. Hopefully the flames of being in love burn down to warm coals of loving that can keep the couple warm for a long time. The evolutionary function of being in love is to form strong bonds of relationship and, in the heterosexual world, this keeps both parties interested in their union for the stability of bringing children into the world.

For those people who find that every relationship ends after a year or two, there is the possibility that they are addicted to the feelings of being in love, the euphoria similar to that produced by drugs, where they experience completion and life makes sense.

Many people experience falling in love with God. As with falling in love with a person, the Ego gets off the throne, and there is the real, lifelong relationship where we make sure daily that He, and not self, is up on the throne. Due to God's eternal nature, the lifestyle is also potentially eternal, as long as we humans hold up our end of the relationship. Worship, prayer and fellowship can all take on aspects of relationship — love songs on the radio turn into love songs about God. Many needs can be met by God's touch into our minds, or He can evaporate the need.

In any successful corporation or organization, there is diversity and a good manager does *not* want all 'yes' people but a variety of talents and energies. This is similar to evolution that likes diversity to play with. So in our quest to attain Sapiens' highest calling, we don't necessarily want everyone to believe the same, we just need Ego to get off the throne. As you can imagine, there would be evolution-abiding Sapiens who would not be able to do this, where personal Ego would have to dominate. Members of a branch of Sapiens might dethrone Ego, but the resultant resident to the cerebral throne would result in a mix of God-believers and nonbelievers, most of whom use science as a tool.

For cooperation and harmony to work, it would have to be that the nonbelievers would not be vehement in their nonbelief, but cognizant that there are phenomena in this universe that are not accessible to all humans, at any one time. Kind of like the Heisenberg Uncertainty Principle applied to people.

Unique Experiences

In the dog world, 0.7 percent watch television and can see and understand what a poster or picture represents.[11] They are able to interpret two-dimensional blobs of shapes into living three-dimensional ones. The other 99.3 percent don't get it. Sapiens who can see can view posters and watch TV, as we have learned through the experience of moving and feeling that our flat vision of the world can be made three-dimensional. For dogs, the 0.7 percent have an extra cognitive ability that doesn't seem to be teachable, and was randomly distributed amongst the breeds and mongrels. Who knows if this particular trait has any evolutionary advantage for dogs?

People can have unique experiences, too, and here are a few examples of these.

Some of us have had the experience of feeling an electric shock in a neck muscle that makes us jump and wince. It only occurs every few months, so we just accept it as one of those quirks of nature where a nerve has randomly fired. When you ask people, "Do you ever get that shock in your neck?", the answer is very clearly yes or no.

Then there is the phenomenon of a cool, sweating sensation in a very discrete dermatome under the eyes, over the cheek bones, when eating vinegar (acetic acid), as with salt and vinegar potato chips. Very few people can relate to this, but again their answer is clearly yes or no.

Here's a funny one that mostly males experience. When finally releasing urine after a protracted retention, the body will shudder involuntarily. Delaying urination is bad for you and if a being cannot urinate for a day or two there is the danger of kidney shutdown and death. So the shudder might be a

11. I did a study on this, printed in the *Canadian Veterinary Journal*, Vol. 48, Jan. 2007, p. 9.

neuronal release, or maybe the closest thing to a spontaneous reward that the body could come up with.

When we go outdoors and are exposed to bright light it will often make us sneeze. During a sneeze it is impossible to keep your eyes open, and so the retina, assaulted by the energy of bright light, recruits a subprogram from the brain to temporarily relieve itself of the discomfort. Dogs do this too. This neurological phenomenon is similar to computer programs that share elements, where a piece of one program is used in another.

Figure 26

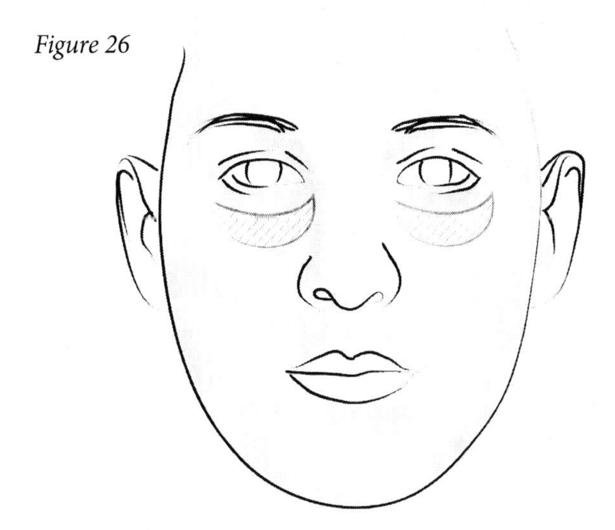

Sweat dermatome stimulated by consuming vinegar

Sometimes while eating, we experience discomfort in our throat, not choking (because it's not in the respiratory tract) but a slowness of a food bolus to get down into the stomach. It can hurt and panic us. It's an esophageal secondary peristaltic wave, commonly known as 'clog'. Clog with cold drink, particularly ice cream or crushed ice, is one of the ways to get 'brain freeze', probably from the cold in the esophagus cooling the carotid artery blood on its way to the brain.

The point is that these variable types of experiences probably exist in every species and the higher the neurological development, the broader the range. There are likely a multitude of abilities that are as yet unknown. It could be that clairvoyance, the ability to tell what's going on in the world without necessarily being there physically, or telepathy, communicating thoughts directly without speaking, are real traits, but for the rest of us 99.3 percent the only way we can know if a person has it is by their success. Of course, the human trait of wanting power would recruit all kinds of charlatans who have no such talent but see the success and want it too. What if having a close relationship with God was like this? It would then stand to reason that some finite small number of people could guide the rest of us on the path to God. This has been humankind's system for ages, whereby a shaman or priest guided the rest of the tribe in issues of spirituality, health, knowing the future and balancing life traditions.

Now, as in the past, those who lack the gift but claim to have it can ruin our confidence in accepting something we don't grasp ourselves. In our recent past, it was the established church that had the scholars who read the Bible and interpreted it for us, but the balance tipped to where oppression outweighed peace and inner contentment, philosophers and scientists questioned everything and people rebelled. Martin Luther led the way with the Protestant reformation of the Christian church, putting the Bible in every person's hands. Now, with neither an authoritative church nor the personal will to decipher the Bible's teachings, many 'Christians' do little or nothing for their spiritual well-being.

Holy Man

When I have a particular problem that I need help with, I seek a specialist. For the bathroom and kitchen, it'd be a plumber; for my heart, a cardiologist; for my computer, a technician or programmer; for my mind and life, a psychologist; and for my soul, a holy person.

It makes perfect scientific sense to seek the counsel of a holy person for a spiritual issue. Sometimes it's difficult to diagnose that it's a spiritual issue that is the core of the problem because the manifestations of the spirit go through the mind, whose workings are described by psychology. Mind and soul have effects on each other but the mind is the spokesperson for both so there can be a conflict of interest. Looking at a holy person's credentials scientifically, you would want to see the following:

1. Consistency — their focus is on the spiritual realm more than this material world, so that their way of being is centred on love more than money and stuff.

2. Evidence of the divine — the fruits of the spirit, or the by-products of living close to divinity, such as patience, kindness, selflessness, consideration and so on, manifest in the holy person, at least more than in us regular folk. Paradoxically, these fruits arise spontaneously the more a person embraces divinity, rather than from concentrated effort. God-believers can think of God as the ultimate divine.

3. Results — their insight into people's needs from a spiritual point of view creates advice and support that the receiver uses to make their life better in a sustainable, long-term way.

Prayer

We use our senses to collect data, and then analyze the data to come to some conclusion about life. Often the data is limited but we come to conclusions anyway. Prayer bypasses all the science of data input and goes directly to the source, reaching into the realm of existence, into the relatively unknown place of spirituality, into the origin and destiny of ourselves, into God. Prayer is the process of removing the ego from the throne of our minds and any other distracting cerebral luggage, to allow the channel to All-Who-Is, or God, to be heard and felt. Most people pray in desperation, when all else fails. This is to say that most people hang onto the current reality, which is an illusion, and this unreality, although stabilizing to a society, doesn't always serve us well. When we give up on that, and pray, we cross over paradigms into godhead. You can live with God on a regular basis and leave behind the illusion of science's promises, going from your mind directly to His being, and gather the benefits daily, instead of just in disasters.

The power of positive thinking can work wonders for our lives, whereby we guide ourselves so that we and our interactions with others are polarized to the positive. Prayer is like that but for those who believe in God, it includes God in the circle of the positive thinkers, and ultimately has Him in the driver's seat.

> Prayer moves faster than the speed of light.
>
> — Christian saying

Like any cerebral function, practice and experience beget growth and maturity, and so it is with prayer. First we start out simply, but the more we grow, the more our prayers seek the bigger picture, as we come to understand that we are a much-loved, essential cog to it — but not at the centre, not on the throne. This is one of the places of contented selflessness where we can experience God. Like the paradigm that Happiness is the default setting that occurs when we remove all those thoughts *we* created that prevent us from being happy, so it is with our relationship with God. By steady practice and growth, we can remove ourselves from getting in the way of God being God, and experience the blessings of completion. This experience can give us a sense that we are participating in the timeless realm. It's one of those activities that you can pursue all your life, into old age, into death, and for all eternity.

Prayer can be one of the most powerful phenomena on Earth, not by the desperate or fanatical, but by regular folk who desire for their molecules, their energy and their spirit to be aligned with the power of the universe.

By the Way

God has no gender — He invented gender. But when worshippers look to God, they need to make a personal connection and that is difficult to do when He is the sun, a mountain or a vaporous cloud in the heavens. From ancient times all the way to recent cultures, it has often been the male who was the boss of the family or tribe and so making the big boss, God, also a male seemed the natural choice. To have a leader God who was genderless would put him outside of the paradigm of humans. It would be more appropriate if we had a term in our language for personal and neuter, but to remove gender would automatically make the being either castrated (less than us) or a freak (not like us). It's interesting to note that many aboriginal and early European religions had a leader God who changed gender or was sometimes genderless, as fit the situation at hand, which is more in keeping with God's omnipotent nature. Being stuck on a Father God is a choice. A logical alternative that is more all-encompassing would be Father-Mother-One or All-Father-Mother. However, at the time when monotheistic religions were getting started there were already religions that arose from animism, the belief of life and divinity in everything, having roots with Mother Earth or Mother Nature, the source of plenty, of reproduction and of food. It would seem that to dominate that spiritual concept, the next order of our spiritual development, also reflected in mythology, was to assign a Father God to the bigger, overhead things like the Sun or the sky.

NonGod

How can a person who doesn't believe in God have any conversation about God? They are being honest and true to themselves, and to pretend to believe would be false. There is no shame in not believing in God. Similarly, the person who believes in God can't see living without Him; how can he unbelieve? The friction arises when either party believes that their way is the only way.

The unbeliever doesn't like to feel like he or she is 'less than' or is inferior because they don't have the perception of the presence of God. Anyone would argue against feeling denigrated by someone else who can't prove the existence of their claim. On the other hand, the nonGod person — in defence, in offence or in retaliation — can strive to make the God-believer feel inferior because the believer follows scientifically unproven concepts.

If people who believe in God were more godly, then there would be an outward manifestation of the inward change, and the world would be in wonder about these people who are truly different in a beautiful way. But godly people live in animal bodies and have animal fears of survival in this sometimes harsh world, and this leads to too many concerns about material needs and things. The result is that the God-believers often appear no different than the nonGod people and so the latter ask, "You can't show me your God and you're no different than us, so why should we believe you?" In addition to this, there are many good people who are ardently seeking a universal Truth that doesn't (at this time) include God. Theology generally acknowledges that a person becomes like the god they worship, and so if someone 'worships' Truth, do they *become* Truth? Many believe that God is Truth, and so if a person attained

Truth they probably would see God, and the other way around too — that if someone goes to God they will know the Truth. Kind of like there are many paths up the mountain, but if we get to the top we all see the same view. Although we might not all agree with this discussion, looking at the billions of Sapiens on Earth right now, most of us can probably agree that what we are doing for our faith walks isn't working.

> If Christians were truly Christian, there probably would be fewer religions.

When you step back and look at Sapiens' history, you can see that those groups that went down the avenue of greater plasticity of mind, through agriculture, better nutrition, commerce, written language, government and armies are no closer to God than those who stayed at various levels along the way. The hunter/gatherer, without science, is in the same position with respect to God as the most highly scientifically advanced person. Science is not the variable. It is another need, apart from the things of this world, that is being met when any mind turns to God.

We Don't Want to Die

> I say that I'm not afraid of death, that I'm ready to go anytime, and add that I just don't want it to hurt. Then I see a dead raccoon at the side of the road and for a moment I *am* the raccoon, living and dying without impact or import on the world or the universe.
>
> Sure, some people love me and life is good, but I don't know anything about my great-grandparents — their names, their personalities, their hopes and fears and accomplishments. My own kin, vapourized into oblivion in three generations, like raccoons, dead at the side of the road. That aspect of me and humanity, our consciousness,

which makes us aware of the bigger picture also haunts us with difficulties dealing with it. One of our biggest fears is that we don't make any difference, that our existence is insignificant.

I was watching my father fold a towel on one of the last days of our trip in Cuba when I caught a glimpse of his existential terror, the motivator for his anger that covered up fear. It was typical of a long quest, where, when you're not particularly looking, it hits you clearly. It was an innocent moment that completed the trip for me and in fact brought our relationship to a happy and full completion too. I told a friend, Ron, about this and he said, in his way of succinct clarity, "What does that (existential terror) look like?" I responded immediately, "Being suspended naked between two planets." In other words: utterly abandoned, a speck in the universe that no one cares about.

We don't want to die, and there are several types of 'death'. Death of the body is our molecular breakup, where we cease to be organized into a sustainable living being. There are automatic programs burnt deep in our DNA, as part of evolution, for every individual to react and respond to staying alive, to keep the body whole and functioning. There is death of the mind — not referring to "brain death" (which would be part of the physical body), but where the mind doesn't serve us well any more. We want to stay sane; we don't want our thoughts blown apart and disconnected, where the mind ceases to function as it was created.

Those fears are unified — death of the body means that our molecules are allowed to move apart into chaos or entropy, and death of the mind means that our thoughts are allowed to move apart into disconnectedness. To counteract these fears, we do everything we can to keep the body whole and to keep the mind cohesive. These efforts all serve to protect and preserve our life on Earth as we know it, but don't necessarily touch upon God or the spiritual realm.

The link across the gap between science and God is something like love, the intangible but real love that encompasses compassion, caring, connection, courage, confidence, wisdom and insight, that makes a person or society whole, and manifests itself as healthy. Love is like God — it has no opposite. Hate is an active emotion of destruction, and not love's opposite. The opposite of love would be something equally intangible like lovelessness or indifference. We can see how they could be discerned as opposites, because one aspect of the relationship between hate and love is that the first works with evolution to destroy and the second works with evolution to support and nurture. Also, they live in the opposite spiritual camps. Sometimes, when survival emotions rise up, it can be that fear creates actions that are the opposite of love. The Devil is not the opposite of God, he is just a fallen angel, one of God's creation who chose to go his own way.

Angel Battle

If there was a battle at the angel level, it likely would be between the archangel Michael and Satan. How the mind of man discovered these things is beyond me, and I have to trust in the holy people whose perceptions in this realm are as specialized as a cardiologist or astrophysicist in their realm.

There is no why or how, just a decision on our part to believe it or not believe it. There is also no application of science, and science will not get you any closer to a decision. It remains a yes or no answer of the heart. As St. Augustine said, "Seek not to understand to believe, but believe to understand." This is totally backwards to science, which demands tangible proof at every step, except in the beginning, at the theoretical level. Interestingly enough, once you make that paradigm jump to believing in

God, you can see that there is no conflict or separation between God and science, but on this side of the leap there could be apparent disparity. Also interesting is that those world-leading scientists who work on the advancing fringe of knowledge, at the theoretical level of scientific thought, sometimes come back to us humble and willing to touch upon the concept of God that doesn't threaten logic. In this regard we see that the more knowledge we have, the more that knowledge will not show us God. God is in a separate department, all His own.

When we look at those groups of people who embrace God and yet live in this world with science's advances, we can see living with a difference. It is rare. There is harmony within the person and within the society, a level of trust that is mysterious and attains cooperative results that secular groups might wonder about. This intangible love goes the extra mile without being asked. It is not afraid of death because it comes from the eternal source and, without limit, flows through us, through life, through death and for all eternity. There is a bridging between us and the infinite and eternal. The vehicle to take us to this nirvana place of mind, body and spirit is *belief*. It's that simple: you either believe or you don't.

The enormously frustrating thing is that even if we believe, we are left with the juggling act of balancing life here on Earth between a disposable body and an eternal spirit, with one foot in the animal world, needing comfort and security, and the other foot in God's world, needing nothing but a relationship with Him. Whenever the animal world grabs us, some of the non-believers point and shout "Aha!" as their proof that the spiritual world is a sham.

Who Needs God?

Whenever there is an imbalance or inequality in the universe, there is an event or a change in forces that arises to restore balance. It could be that the Big Bang or DNA (the Little Bang) arose this way, after which time the new creation followed the laws of physics and nature. The greatest mystery at this time is what caused the new beginnings to come into being. For now, those in the God camp easily allow that God caused it, but some of those in the nonGod camp won't want to abandon their belief that Science will eventually explain everything.

Once we have answered all the scientific questions of the universe, and shown without doubt how the first molecule of matter came into being, how the universe has grown, by a universal simple mathematical equation, we will still be existentially where we are now: without an answer to the question *why*.

It is the crucial question of why that is the core of the meaning of life. It is okay if you don't need meaning at this time, but most of us *do* need it. This is not to battle the insignificance of our entire species, but to solidify our place in the world. If we can see the future and control it, if we can enjoy mastery of the space-time-matter-energy-light continuum and claim ourselves God, we will still bleed and cannot be God.

Heaven and Hell

'Animal' is what we see and feel right now: warm or cold, hungry or full, sexual satiation, ambition or contentment. 'God' leaves the containment of molecules behind and takes us into the limitless possibilities of completion, fulfillment and purpose. 'Sacred' is the interface between us Sapiens, in animal bodies, looking to bodiless God, and getting a small taste of eternity, answers and peace.

To the God-believers, Heaven then is life with God, in His realm, and Hell is life without God, and so Heaven and Hell aren't places somewhere or sometime else, but here and now, and eternally and everywhere we are. The less we are with God, the more we have fears of existence, of insignificance, of meaninglessness in spite of our best efforts. Much of our anxiety and ambition is due to the effort we make to alleviate this fear, often without our awareness of the battle.

Who or what do you believe in?

- The Earth
- Your continent
- Your country
- Your county, province or state
- Your city
- Your neighbourhood
- Your block
- Your house
- Your family — spouse, kids, parents
- A star of music, movies, athletics or politics
- Your favourite sport or team
- An ideal, like peace on Earth
  or scorching the Earth
- Your church
- Work
- Money
- God

Could it be that our peace lies in understanding that our lives have no meaning in the material universe as far as our molecules are concerned, but that who we are as beings has meaning here and now, in the way we relate to each other? The God group would add, "You matter to God." Or, put in another way, you are loved and your life has meaning for all eternity. Or is it that the worst fear, "being suspended naked between two planets" is actually a view to our salvation, and knowing that we are, both molecules and energy, essential cogs in the universe.

In some sense, if one of us is missing from our realm, the universe is incomplete. If we originated as one single cell in our mother's womb, and all life on Earth originated as one single cell, and all planets and stars originated out of one single explosion of neutrons, that might have originated out of pure energy, $E=mc^2$, then, really, all life and matter are related.

If, as scientists, we consider that everything originated out of nothing, why should we be afraid of the Nothing: no meaning, no future, no God? As the yogi said, "Can you stop death? Then why be afraid of what you cannot stop?" It's because, unseen and undiscovered, there is an energy or drive within us that strives to make meaning out of life.

As I wrote this, the radio played George Harrison:

My sweet Lord,
Really want to know you,
Really want to feel you,
Really want to see you, Lord,
But it takes so long,
My Lord.

The Lava Lamp of Religions

A Jewish man named Abraham migrated north from the most advanced city on Earth at the time, Ur, Mesopotamia, to the Mediterranean coast, and his son Isaac gave birth to the line that leads to Jesus and his son Ishmael gave birth to another line that leads to Mohammed. In those two offspring religions, Christianity and Islam, there have been several splits.

In the Christian world, there are orthodox churches, like the Greek and the Russian, that don't subscribe to Roman Catholic rule, and apart from these, there are 2,000 different protestant churches that have arisen since Martin Luther and the Reformation: United, Presbyterian, Methodist, Baptist, Reformed, Lutheran, Evangelical, Anglican, Episcopal, Gospel, Pentecostal, Mennonite, Foursquare, Seventh-Day Adventist — to name a few. From the protestant group have also arisen the Jehovah's Witnesses and the Mormon faiths. The religious lava lamp has been very busy.

In the Islamic world, the Shiite and the Sunni factions had a falling out and forgiveness and reconciliation do not seem forthcoming. The Jewish faith doesn't give us much scandal — they have the stricter Orthodox, the mainstream Conservative, and the easygoing Reform, but they don't seem to argue much about their belief system. The most peculiar aspect to Judaism is that, although adherents total less than 1 percent of the world's population, the Christian world pays so much attention to that part of the world, because of it being the source of the Christian religion, with Jesus having been born a Jew. You have to wonder, if Jesus had been born of Saxon, Celtic or Roman blood, would today's news show us as much as it does about the Middle East?

We often think of cultures, races and languages as distinct and separate, due to our short timespan of observation, which would be our lifetimes, or a few generations or a few thousand years. In actual fact, everything is in the process of morphing into something new, either by evolution within itself or by amalgamation with/engulfment by others or division of self.

Figure 27

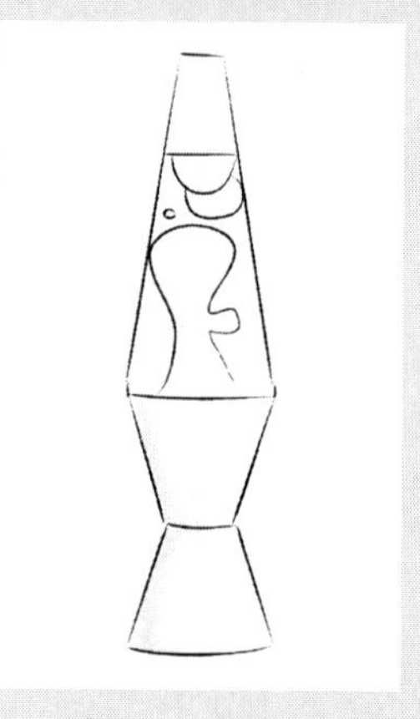

Lava lamp

When you look at a lava lamp — one of those tall cylinders of coloured fluid with gobs of viscid semi-

> liquid rising, merging, then falling, over and over — and imagine the gobs being different colours, you'll get a picture of the process of most of man's development on Earth. One aspect of Sapiens, such as language, rises in the lava lamp as one coloured gob of liquid, merging with another coloured gob, and the two become one as the resultant new colour. One colour could engulf the other but the two don't mix and can be seen to be distinct still. The same is true for races or ethnic groups.

There are other religions of the world — close to a billion people follow Hinduism, about a billion follow Buddhism, local tribal and folk believers add up to more than half a billion people and there are variable millions in each of Baha'i, Jainism, Sikhism, Shinto, Taoism, Wicca and Paganism, Zoroastrianism, Confucianism and new religions. Avowed atheists amount to just 2 percent of the world's population, but there are growing upwards of a billion people who have no set beliefs, or who are part of a discernible but unorganized group such as agnostics and humanists.

The point here is to observe and understand that nearly everyone has thought about spirit and destiny, that there is no uniform agreement and that any one person lives in a paradigm that works in their neighbourhood but is probably not shared by most of the rest of the world. So we agree in principle that there is a spiritual world out there, but there is little agreement on what that looks like and our relationship to it. Termites and cockroaches show better cooperation but they are not blessed and plagued by a higher consciousness (that we know of).

All of these belief systems have arisen in the last 5,000 years, except folk and tribal faiths which probably have been around since the cave paintings 30,000 years ago. We've been around as a species for 200,000 years and *Homo* has been around for maybe 2 million years, so you can see that our awareness of God and the spiritual world is very recent for us, a mere blink in the span of our existence. Although all these different religions contribute to a huge variability that social evolution will use, there could also be some aspect of physical evolution as well — in a global crisis, this could play into selection of certain groups to survive while others perish. Ironically, this only involves molecules and the material world, our bodies, and does not address spiritual survival. Once again, after being open-minded and discerning all the variables, we come to a humble place where we cannot prove the best course of action. It resides in faith.

Has God Evolved to Suit Us?

Not really. All of us humans, all over the world, became aware of the spiritual world well before the cave paintings were created in Spain and France more than 30,000 years ago. Life was and is full of mystery, and prehistoric people would have struggled with knowing that there was more going on behind the scenes of life, and not quite knowing how to deal with it. Everything eventually acquired a spiritual side — the Sun, the animal just killed for food, a tree whose bark was harvested, a rock, a river and, of course, us. We have seen through the analysis of life down to the molecular level that the material world cannot explain everything that goes on in the universe, that molecules are recycled through everything and that they could well have memory of their own.

This oneness of the Earth was likely intuitively grasped by the ancients, and it created the awareness of the spirituality attached to everything. They lived as though everything was connected to divinity at that moment. This, on the one hand, is a marvellous ability rather like praying without ceasing while going about our daily routines, a talent that most of us have lost. On the other hand, there would be difficulties separating material world activities from spiritual ideas where some spirit or god would not be pleased with virtually every action, unless some ritual or prayer was invoked. This could even go so far as to interfere with survival. As was the case with farming, where equally intelligent people chose to remain in the nomad-hunter lifestyle or chose to take on the stationary and more complex life of farming, so it was with the type of spirituality. Some people chose and some still choose to live their spirituality as the ancients did, while others moved on to develop gods as a way of organizing their spiritual ideas.

The Celts were probably the last vestiges of a spiritual culture within the Eurocentric world, the white man's aboriginal people. Although there are other aboriginal Europeans, such as the Sámi in northern Scandinavia and the Basques in France and Spain, it is Celtic music, dress, art, dance and language that is attractive to many people, as it nudges our dormant spirituality and sparks a yearning for a missing part of our lives. Yet, many of us still return to our need of power, money and science, bereft of deeper satisfaction.

We're aware of the Greeks', Romans' and Vikings' gods, who were personified and had job titles. The same idea pervaded in other parts of the world — China, India and in the Americas. There were lesser and greater gods, but they usually had a personality that we could relate to — we anthropomorphized them to make the conversation easier. And there was usually the boss god, one who was mightier than all the rest.

Then came the idea that there was only one god, God, and the religion that we are most familiar with as starting this monotheism is Judaism. God was the source of all, eternal and before time, was in all and of all. This description amazingly captures the essence of the spirituality of the ancients, where divinity was in everything, except that this god was in control of everything, loved his creation, especially us, and you could argue with him.

The mind of Sapiens had developed over the last few thousand years, and society had grown up to be more tolerant and intelligent about the way people treat each other. This social maturity was reflected in the Old Testament. It would seem that by pre-Old Testament/Judaism if someone hurt you or took your stuff you could kill him, because the new rule limited your actions to "an eye for an eye and a tooth for a tooth" — not murder. It was a new approach, to get your justice or revenge in the same measure as the crime.

Two thousand years later, Jesus Christ came along with another major paradigm change — don't fight at all but "turn the other cheek", and if a guy wants your coat, give him your shirt too. He said 'Don't be afraid' in many ways, many times, because he knew the state of humankind, he knew about evolution and that we are driven by the fear of not surviving. He said to forgive "not seven but seventy-seven times", showing us that forgiveness is one of the most powerful tools of love. This has heavy psychological implications as well, for many people develop pathos from blocking forgiveness, either giving or receiving. It impacts international relationships, too, as countries can act like individuals and withhold forgiveness. Jesus tried to tell us that the path that we are on will not lead to satisfaction or happiness, and that focusing on the eternal can make our present moment more complete.

The Mayans have their calendar, the Muslims theirs, the Hindus theirs, and there are others, but most of our modern world has a dateline based on the life of Christ. Some early Christians were killed, some ran, some argued their case, but the end result is that enough of them survived to continue the faith. There had to be some huge driving force that kept them going when so much was against them, and it was their experience in knowing that Jesus opened their eyes to God.

Evolutionary Effect of a Belief in God

There are scientists who believe that belief in God and the spiritual world was an asset in survival, maybe by assuaging fears of the unknown or fear of mortality. We in the West are surrounded today by so much technology and organized civilization that many people feel so safe in the day-to-day care of their molecules and cells that there is a huge postponing of thoughts of mortality. Science and medicine promise all the answers and immortality, so when someone gets sick and goes into a big building to die, we have a ceremony and tea party, then quickly get back to our illusion of life without morbidity and mortality. Without this enormous construct that society has built you can be sure that people would once again be thinking about the spiritual world and God. What this means is that our successful living is part of what is separating us from the spiritual.

Just as physicists are looking for a unifying theory that can describe all phenomena from subatomic to celestial, it might be gratifying to find a unifying principle that would tie together this world and the spiritual world. The bridge could have been proposed as energy, but because the spiritual world steps outside of time and space, which are the home of energy, there would be no link.

As incredible as the power of energy can be in the material universe, it is a *finite* entity. As far as we know, God and His realm know no limits — as we step into His realm as spiritual beings, we must then have no physical limits either. If we still had limits, like energy, and stepped into His realm, we would bring with us some kind of limitation, making limits within His realm, which would be incongruous.

Also, if energy was a shared commodity between this material world and the spiritual realm, then the spirits could cause effects amongst us and that would be a common experience, which it is not. Some of us may have had or may have heard of someone having an experience that suggests some kind of spiritual effect on us or our surroundings, but ghosts and spirits moving our material world isn't an everyday occurrence. There is the aspect of miracles, which is God directly making material changes amongst us or within us, but how He does that wouldn't necessarily be expedited by energy.

So here we sit, with no connection between Quantum and Newtonian, and no *measurable* bridge between material and spiritual. Jesus himself existed on the molecular level and is now spiritually eternal, and showed us that we can do it too, simply by believing Him, but the measure would be inside us.

Applying the sciences of sociology and psychology to people, we can see the effect of God by the changes in a person as they grow closer to God. Sometimes the changes are practical results from environmental manipulation or changes in how people think. Often the changes come about when a person accepts God in their being (in their mind, in their heart, in their soul) — God, transcendent of molecules and energy, God or All-Who-Is, God personal, not ethereal or theoretical, God who knows you and loves you, from before you were born and into all eternity. That step is one gateway through which all other spiritual phenomena can follow.

Sapiens Didn't Invent Spirituality

Most tribes of Sapiens have members who believe in some kind of god and spirituality. It would appear that Neanderthal in their latter stages buried their dead with their personal possessions, likely indicating that they thought that the dead person might be needing those things after death. If they were simple cavemen, any article like an ornament or tool would be up for grabs as soon as its owner was not capable of guarding it.

This shows Neanderthal was a sentient being, also with consciousness — a type of human. In this way, Neanderthal acts like a control group, which documents that beings with consciousness need God, or a relationship with Him. Perhaps Neanderthal was a practice species,[12] a warming-up to a higher consciousness, preparing the way for Sapiens, and once again showing us Nature's or God's mysterious path for Life to become more organized and complex. Scientists might argue that Neanderthal, like Sapiens, had similar brain anatomy and

12. We are learning more about the small overlap between Sapiens' and Neanderthal's genomes, more likely from common ancestry than significant interbreeding.

therefore physiology, so after attaining consciousness had a similar evolutionary need to believe in God. This would beg the question to scientists of how many believer species would be needed before we accept their testimony for the existence of God? If every species on Earth bowed and prayed in the same accord yet scientists had no measure for God, would they cling to their recording instruments and say that He doesn't exist? The ancients lived with the divine in everything, even the inanimate, proposing by lifestyle that all life can believe, and so it could be that Sapiens, with the most ability to conceive through his higher consciousness, polarizes to the opposite view as well, that of unbelief.

Alternate Realities

In any society there are conventions as to the accepted ways of living together. There are many ways, cultures, rituals and styles of societies, and these are realities that are partly real and partly invented. Each reality bumps against unreality or alternate realities, and the two are separate but connected. Just how far we venture into unreality depends on the culture. In Western culture, we adhere to our perceived scientific reality to the extent that we do not approve of going into unreality. We are very careful and critical of what is real and what is not.

At the same time, occasionally there is an artist who uses paint, the written word, film, sculpture or music to draw us into the fringe of alternate realities. When we regard a painting by Van Gogh and see that the objects appear to be shown from different perspectives, it's a little crazy. If we could ask Vincent why he painted that way, he would probably respond, "What way?" Picasso presented us with Cubism, trying to make us see ourselves in a different way. Every generation has a music style that is mostly repugnant to the older generation, but after a time and a melding of the new style into mainstream music, they all become accepted. These artists push us out of our comfort zone with new ways of defining reality.

In aboriginal cultures, there has been the induction of alternate realities, both through drug-induced states and non-drug methods such as chanting and dancing or deprivation of sleep, food, water or comfort. There were always the specialists, or shamans, who were more experienced at this. They would venture into their minds, seeking the netherworld, other world or spiritual world in order to learn more about life in all its depth and breadth. They would come back to reality and report to the rest of the tribe what they had found. Today, people view these alternate reality experiences as simply drug-induced hallucinations, or a bit of craziness, to be avoided. Yet going into these alternate realities must have served a universal function in the survival of our species, because people the world over have sought them.

As we have become more modern and scientific, we have left the shaman practices behind and have only the artists and physicists to take us to the fringe of reality. We commonly partake in drug-induced states with alcohol and drugs, mostly to make us feel good, but you have to wonder whether these states are part of a yearning to be more complete, closer to God, or some intangible drive to seek an alternate reality. If reality was all we needed or wanted, we would sit in that and be content.

The Conundrum

At every level of understanding, we can grasp what we know and beyond that point some would say that anything unknown is due to God. Knowing God isn't about giving Him what is leftover once Science is finished. God is not what is left over nor is He responsible for the leftovers. Knowing God has its own path, and it is a misunderstanding to allow our current understanding to determine what He is or isn't. The decision for God is always with us, at any level of logical understanding.

With all these myriad intangibles, it is perplexing to know what path to take, how to live and what to believe. Both Nature and our society would have it that all types of beliefs should exist together, both the God and the nonGod peoples. Physical evolution doesn't really care about individuals, just about gene pools, and so that side of us would want to struggle to be all that we can in order to pass our genes on to the next generation. So we are left with the questions of what is best for us individually, for us as a group, as a species and in our relationship with God, and whether there is any link between all of those answers. There are no easy answers. The answers lie in the zone between black and white, where we swim in the greyness. Awareness is likely the key to finding our place, and possibly the path to our species becoming able to choose how to evolve and how to be.

9

Evolution

Origins of Life

We should start a conversation about evolution with some definitions to offer clarity instead of controversy. Evolution has never been presented as an explanation for the origin of life — they are two separate issues. Living beings change genetically over generations, and when challenged by a change in their environment, the new circumstances will favour some and deter others. This is evolution and it happens all the time. The scientific community and most theologians agree on this. As for the *origin* of life, science has several theories that attempt to explain it, with nothing proven yet, and some religions have ideas about it that don't necessarily ever need to be proven, as they are allegories meant to give solace to the soul more than data for the brain. In this regard, the stories from either party can exist side-by-side, as one fulfills our desire to know the universe and the other allows us to pay homage to our existence. Controversy arises when one party insists on 'being right'.

God is not threatened by science or theories. For if God is so fantastic to be beyond words and comprehension, and is — if you believe it — the maker of the universe, then his power is greater than our microscopes and telescopes.

> But we, the maker of theories, need to question whether or not we want the mind to be the sole spokesperson for the soul.

Most of us have no opinion on how life started because no one has given us a provable answer and, besides not wanting to look foolish with a potentially wrong idea, those types of contemplations don't help us get on with our day. While most of us haven't come to any conclusion as to how life started on Earth, some of us intuitively roll the clock back, reversing evolution back to progressively more simple living creatures. At some miniscule size, we end up at a single cell and, continuing with our reversal, break the cell apart into vital component parts. Now when we run the show forward, we see molecules assemble into coherent parts, DNA come into being, and a membrane wrap around the collection of molecules until a primitive cell is born. From here, it's easy to imagine evolution working on this first cell or cells, that over billions of years eventually resulted in more complex living things.

Figure 28

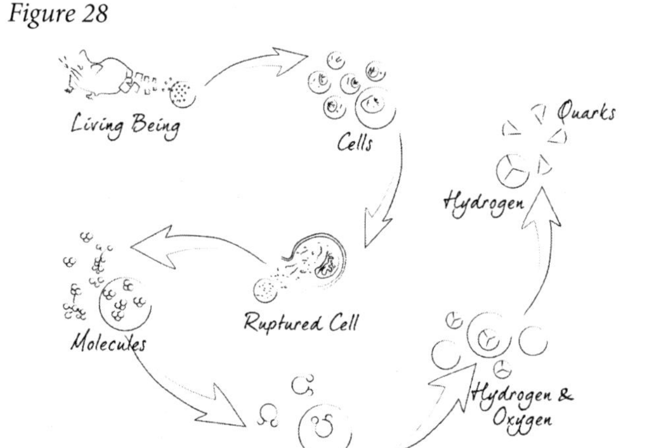

At what point do the building blocks of life actually become life?

While there are many legends, myths and stories around the world as to how we got here, how life began and how humankind came to dominate, here are two examples of the more commonly known ones.

One legend of our beginnings says that billions of years ago, the planet Earth was young and there were all kinds of chemicals abounding, water to slosh it all around, volcanic heat from the Earth's still partly molten crust, thunder and lightning, and the sun's raw energy blasting through to an Earth without atmosphere. There were essentially no limits on timeframe. This warm, energetic, perpetual soup played with statistical encounters of all kinds of chemicals in infinite combinations, until one day DNA came into existence. It didn't break apart, it stayed. In spite of all the other chemicals coming together and breaking apart, DNA was a winner — it stayed. For an encore, it doubled itself perfectly, to make two identical helices, entwined together to make the famous double helix. It somehow developed the ability to produce complex proteins, to assemble fats and to orchestrate the manufacture of a cell wall to keep all other chemicals out until they were invited. And there you had it — the beginnings of life. This is known as abiogenesis, meaning 'no life beginning' or non-life starting life.

From there, with loads of time to experiment with different ways of doing things, some cells survived better than others, depending on the surrounding conditions of heat, light and which chemicals were available. In that process, these first cells had to work out the tricky business of reproduction by equitably sharing their cellular soup between two newly formed double-helix DNAs, so that the new cells were exact replicas. Then at some point, two new cells somehow agreed to stick together, instead of being free-living entities. Then everybody was doing it, sticking together. This was taking millions of years. Once the cells successfully stuck together, the statistical experiment continued, trying this way and that way of doing things and of being. Eventually, some cells were doing certain things in a more specialized manner so that the bigger clump of cells could get along and attain some mutually beneficial goal, like surviving intact and successfully reproducing the whole entire mush of cells into an identical mush of cells. Add millions of years of this and you have *Homo sapiens*. This legend combines the story of abiogenesis with the story of classical evolution, and so the complete story could be called abioevolution, or evolution that extends back into raw chemicals to explain how we got here.

There is another legend that is written in spiritual belief systems around the world, with variations on the theme, and also found in familiar books — the Jewish Torah and Christian Bible. It says that God, an unseen, intangible but real force, took raw materials from the Earth, clumped them together and breathed life into it so that a man came into being. Then, in the biblical story, God plucked out a rib from this first guy and used that to create the first woman. These two original people had children who became the human race or *Homo sapiens*. This legend is called the origin of life by creation.

Figure 29

Ants arguing

Some people argue about which legend is correct, which is like two ants wrestling with a morsel of food while, unbeknownst to them, a bulldozer is approaching.

We see that the first legend comes from the perspective of understanding the universe by the process of systematic thought based on the information at hand. It tackles the riddles of eternity and infinity with courage. The second legend comes from the perspective that *our relationship* to infinity and eternity is just as important as the existence of those mind-stretching concepts. One way of humankind having a relationship with infinity and eternity is through God, or, as there are some belief systems that do not have a god, spirituality. Science uncovers the mysteries of the universe and through our relationship with God, we discover our place in it. There are many scientists in the world who are deeply spiritual and there are theologians who are keenly interested in science.

At this time, no one knows the origins of life. The power of science is in its potential ability to analyze how the first molecules became life, and the power of spirituality is in its ethereal ability to tell us of the origin and purpose of our spirits. As we are made of both molecules and spirit, the cooperation between the two disciplines might give us a complete picture of our existence.

Carbon Laughter

I was talking to my friend Rob about all this, and what it boiled down to was that he *believed* that random creation of life was possible. He believed the 'evolution' legend could explain the origin of life. But for a scientist to back his argument with belief is breaking the rules of science. Science is based on fact. So when science does not come up with a satisfactory explanation to a difficult question, it could be that it stimulates our fear of the unknown. This is not to encourage leaping to the concept of 'God did it' to diminish our fears of the unknown beginnings. When my friend and I carried on our conversation of the origins of life, we looked at the chemicals necessary for life to start, in particular carbon, the backbone of life. In the end, he admitted that he didn't know where the carbon came from, which sent us both into fits of laughter. It brought to the fore the question before life — where did the building blocks, the chemicals, come from? One of science's answers would be 'the Big Bang' and some God-believers would say "God made them." Our laughter told us that we were on the right track — the most important thing was our love for each other, spending time together and a good meal, not being 'right'.

Long Live Evolution?

The most important and somewhat chilling concept to comprehend about evolution is that the surviving generation is not necessarily the most talented in speed, strength or intelligence, but in its ability to *adapt* to the changing circumstances. We could keep that in mind when our circumstances change — our flexibility and versatility are our greatest assets in rapidly changing environments, which currently, for Sapiens, is social change.

Evolution cannot explain the origins of life and doesn't try to. When you roll time backwards and get back to the first cell on Earth, there is a gateway to go through. Molecules and atoms are not life — they have consistent quantum masses and energies that make them entities, without abiding by the current definitions of life. Therein lies an exciting concept that we might not know yet, that life has a broader definition than we acknowledge. For atoms to leave randomness, entropy and pure forces and assemble themselves into a living cell was a big step. Perhaps it was many tiny steps, but with an intent beyond randomness. As sure as x divided by 0 has no meaning, randomness forever is just randomness forever.

Abiogenesis theory might one day explain how molecules became life, but likely still will leave us with the unanswered question of why. We know that we do not know how life started. That leaves us everyday people with no theory, which makes us feel that we're not in control of things and that can be frightening. So we fill in the void with concepts, which in some ways is a good thing, as it will eventually be filled with the truth. In the meantime, we need to be clear that evolution in itself does not try to explain the *origins* of life.

There are different Christian interpretations of the Bible's story of creation — from fundamental and exact acceptance of the story through to more allegorical views. Most of those viewpoints are enlightened by the knowledge that the story has some overlaps with science's observations. The creation of the plants and animals as outlined in the Bible would approximate the origins of life.

What is fascinating is that the writers of the Old Testament, about 3,500 years ago, describing how the world was created in seven days, tell that plants were made on the third day, fish and birds on the fifth day and animals on the sixth day, exactly in the order that modern science believes that they came into existence. It gives credibility to thought processes outside of science.

Ascribing some Being as the originator might not be so abstract as you'd think, but the scientific mind would ask, "Okay, but who made the Being?" This is a fair question that we shouldn't be afraid of, with science in our hands and, if we choose, God in our hearts and souls.

Further to that, the big event that we're seeking, the attainment of consciousness, is described in the Bible when Adam and Eve were created — Man came into being, leaving the animal world behind, and in the process his mind opened to the awareness of God.

Another possible analogy would be that Adam's creation describes hominids coming into being, and that the bite of the apple was the event where consciousness arose. The key concept is that it took a moment to achieve by either description: God created Man, or Sapiens' mind was opened.

Life Here, Life Elsewhere

If you believe in the creation of life by 'evolution' then you have to agree that there is life on other planets. It's not reasonable to have a theory for this planet and then deny it for other planets. With zillions of planets out there, and all the time in the universe, if random chance describes how life arose, then statistically it's a given that by random chance life could occur elsewhere, a thousand times over.

Figure 30

Image courtesy of: HST/NASA/ESA

The Sombrero Galaxy
has fifty million stars,
each possibly with several planets –
so one galaxy alone shows
the potential for other Earths

On our planet, with our system of evolution, the most advanced beings are the ones highest on the food chain, the carnivores. Somehow the hunters and killers have more complex thought processes and actions. So if our neighbours on other planets are more evolved *in the same system* as we have, then it's possible that creatures from other planets could be far more aggressive, and therefore pretty nasty.

If we admit for a moment that some UFOs (the 'real' ones) are glimpses of visitors from elsewhere, then they've attained advanced forms of space and/or time travel. In order to have a society that got to that point, they'd have to have evolved in a different system than us, or found a way not to destroy each other, as our Sapiens societies always do when we get to a certain point of advancement.

Aggression blows things apart, intelligence brings things together. Any advanced alien culture would probably have come to the same crossroads that we are at now — decide whether we are going to continue to let evolution (unlimited competition) be the guide of our destiny, or choose for ourselves. This, as Professor Hawking has already warned, might add another element of caution for us attempting contact with aliens.

Further to that, if we step outside our paradigm for a moment, we as a species might not even be on their evolutionary tree. Or if we are, we might be embarrassingly lower down, not even as a lower primate but maybe something more like a worm. Wouldn't that crush our arrogant species' ego? Christians and Muslims might get upset by this analogy, but Buddhists probably wouldn't, as they see all life as precious, including a worm.

However, being irrelevant to other superior beings could be our saving grace. The ultimate power in the universe isn't the ownership of land, money, gold or scarce material goods, but of access to and control of *energy*, so we don't need to worry about them farming us for food.

Nutrients are building blocks that are abundant in the universe, and food is one source of this that is also a form of energy, captured from the sun and converted into usable form for us. For a civilization that has mastery over the energy of the universe, energy in the form of food would be very simple to attain, so we should be very safe from consumption.

What Next?

One possible progression that could occur within Sapiens is for a next big event or gateway to occur, where another Adam sees something that has always been with us, but is able to grasp it and show it to the rest of us, some of whom will also grasp it.

For example, it could be that someone, then their group, will develop a means of communicating that will be exponentially faster, like speaking with the rapidity and precision that computers have electronically. There are different abilities and talents within our species, some of them lying undiscovered so far. If a select group, maybe the 0.7 percent (like the dogs with extra cognition), find a transcendent way of being, by clairvoyance, levitation, telepathy, thought control or whatever, their survival will depend on relative isolation from the fearful mainstream of humanity. 'Relative' because these evolutionarily advanced survivors could walk amongst us but have a safeguard of some kind, rather than needing to be physically isolated on a separate continent.

Change Is Coming

One of the aspects of evolution is that it relies on genetic drift gradually changing the internal makeup of individuals and therefore species, and then a stressor coming along that tests all the various models of the species at that time. Sometimes the stressor force is cataclysmic and wipes out the species, other times it favours one type or model within the species that becomes the sole surviving part of the species. So it is with societies and cultures — they are constantly in flux, to different degrees in different areas, and make changes slowly until a major stressor comes along to test their way of being. Often it is at the hands of war (intraspecies competition), but could also be from famines and water shortages or pestilent disease. There is occasionally a massive volcanic eruption, earthquake, tsunami or asteroid that ruins the environment that Sapiens needs to live in.

With so many people on Earth now, the stage is set as it has never been before, as we have far overstepped the boundaries that would naturally contain us. There are so many vulnerabilities in our species' existence that any one of a number of catastrophic possibilities could arise to rebalance Earth's life equation. For example, it might not be a virus against *us* but maybe a virus against something we need, like wheat — as John Christopher describes in his novel *No Blade of Grass.*

In our speculations of future ways and reasons that the worldwide population of Sapiens could drastically diminish, there isn't often a positive process offered, such as voluntary control or cooperative management. One way could begin with a thought or a *new consciousness*, as radical as the original creation of consciousness, that through neuroplasticity or some as-yet-unknown process, we have our minds opened to fantastic possibilities. We are definitely ready for a new paradigm of existence. We might have to go through some rings of fire to get there, but we have to imagine that our destiny is something more than survival. Something wonderful.

Some of us are ready.

The Choice of Our Future

What path do we take so that we are using our brains to choose how our lives and species should evolve? That's a tricky question because no one knows what pressures will be exerted on the species of Sapiens, and so we have no prediction of what the adapting survivors will look like. We can go with Nature and the animal side of ourselves, which is currently leading to catastrophe, the aftermath of which will not see us as populous and advanced as we are now. Or we can choose how we want to be, in principle, and let the floodgates of our development open in that way.

The negative aspect of our animal side, with its fear, anger, jealousy, greed/avarice, domination and all those nasty aspects of the survival game, cannot support a culture as advanced as that depicted in *Star Trek*. Disagreement and war will always threaten to annihilate us, all the more so as the power of the weapons escalates. Someone once said that they didn't know what World War III would look like, but that World War IV would be fought with sticks and stones. So it appears that at the present time our species is not mature enough yet to handle the greater powers of the galaxy.

As the power of the mind is our strongest evolutionary asset, that is the place that we need to develop and grow, but the mind expanding without spirit could be just another forum for evolution to exert its powers over us. Until a better way is found, a possible path for our successful future lies in the union between intelligence and compassion. We need to open the gateway that allows science and technology to walk hand-in-hand with spirituality/love/God, and let that be the path for our thinking processes. Without this balance, more advanced thoughts in an animal brain could result in weapons, drugs and sex on mega-steroids, and the outcome of that is too predictable.

New Species of Humans

We see that there is a certain proportion of people who are born generous, some born for the chase of athletics, some natural leaders, and so on. Were there any serial killers when we lived in small towns, before large cities developed? Or has this phenomenon arisen out of the well-known 'too many rats in the cage' observation? (Their fertility decreased and cannibalism increased.)

We look back at all the different Homo species and attempt to unravel the mysteries of who lived and how. Yet we who live now are in the middle of what will be history and it's possible that there are different species of Homo now living side-by-side, but we can't tell who they are because there hasn't been a big test yet. We might be splitting into *more than* two new subspecies, but here is one proposed idea based on a simple branching of our part of the species tree:

HOMO NATURALIS is that portion of Sapiens that clings to the old ways, where the natural way of evolutionary competition is still the practiced way of being, complete, when necessary, with aggression, greed and violence. Naturalis societies can be anywhere but when they arise from large city centres they create billionaires and armies. If a big stressor tests the genetic and social viability of a culture, Naturalis is likely to survive, *if he doesn't kill everyone*, but the survivors will likely have to live as we did in prehistory. These people are fully engaged in the struggle that evolution brings. Changes in the brain can be occurring from repeated environmental stimuli, reinforcing forcefulness as a way of being, and not just from learned behaviour. Strong negative influences on children can change their gene function, which illustrates that our current society can create a subsequent one that is not different just from fashion, but that their being changes. These types of factors could possibly contribute to the biochemical changes that promote a new species.

HOMO PHOENIX are those Sapiens who would ascribe to the idea of choice as a way of being, where we take care of ourselves but also our communities, and competition creates excellence without leaving others downtrodden. These are the people who strive to learn the secrets of the universe and to harness the power of the galaxies for the maximum pleasure and fulfillment of all members of society. Neuroplasticity supersedes evolution, and they might develop and use their brains and bodies in newer ways that allow for increased awareness and intelligence. If this group attained the abilities of space and time travel, it could make resources limitless and make money obsolete.

HOMO SAPIENS remain the bulk of us, the reserve pool from which the other species are arising. We see indications of Naturalis and Phoenix at work, but have little inclination to change to anything other than what we're presently doing; we are the great inertia. We strive for family security and common decency, we live in a huge mix of cultures, religions, races and languages that form the crucible of humanity, at once accepting and intolerant, loving and hating.

In all three of these subspecies, we will find the spiritual and nonspiritual, peoples of aboriginal and developed worlds, artists and scientists. The new species will not be determined by who we are but how we adapt, and, for Phoenix, also that we agree that nature is now turning against us and could be a force unto itself that we now need to harness and guide.

Alternatively, it could be that Sapiens has always been the wild one and he is becoming irrepressible through crowding, successful reproduction and social pressure, so Naturalis is really just Sapiens unleashed. To round out this second viewpoint, Phoenix could be quietly evolving off to the side, having its brain develop differently in yet another way, due to *its* chosen path.

Two species of Homo could live side-by-side, as with Sapiens and Neanderthal, but if Naturalis is the aggression specialist and Phoenix the cooperator, the latter would have to find a self-protective mechanism for safety without violence in self-defence. Perhaps larger numbers of good old Sapiens could form a protective buffer that dilutes Naturalis' effects and allows Phoenix respite in the mix of humanity. As mentioned, there is this unspoken assumption that only one Homo species will survive, when there could be several hominid species living simultaneously.

This presents a third viewpoint that there could eventually be as many separate genetic species of Homo as there are now separate races of humankind, to act as a launch pad for the next evolution or choice, and to be the mixing pot and reserve pool from which that next step comes from.

What will our successor hominids look like? Be like?

As an illustration of how the forces within evolution and the possibilities of choice can work, the National Geographic video *Stress: Portrait of a Killer* shows a group of baboons whose most aggressive males dominated over a tainted food supply and died from infection (tuberculosis). More aggression, more food, more infection, more death. This is similar to some of our modern societies where food abundance is causing medical problems such as obesity, diabetes, heart disease and death. More success, more food, more death. The baboons left over were more peaceful and got along fine without the aggressive ones. Moreover, when a random baboon came along looking for a new troop to live in, and approached the group with aggression, the residents didn't respond in kind, but kept to the peaceful ways. He learned to be peaceful like them and gave up the aggressive stance.

The point is that Phoenix might somehow convince Naturalis or Sapiens of a better path, maintaining a larger gene pool through neuroplasticity. To put it another way, it might come to pass that part of our species adapts to a new way, then teaches the rest of us. Those that follow in adaptation will survive and those that don't will not. It could involve a selection process that doesn't include catastrophe — perhaps more like attrition.

10
Conclusions

Is LIFE ON EARTH a one-trick pony — evolution — or do we have a choice?

Self-control is the key to achieving goals. The athlete controls his body to attain achievements that the rest of us can only dream of. Scholars and researchers control their thought processes to gain knowledge and clarity to share with the rest of us mortals. And regular citizens like us control our morality and urges to maintain the objectives of living together in harmony — our mandate is the self-control required to avoid the calamities caused by theft, adultery, lying, judging, criticizing, dominating, controlling, blaming, and verbal and physical assault.

The purpose of every generation is to provide the garden from which the next one grows. There are also generations of humankind, where occasionally some individual or group comes along that transcends the previous one, taking us all through a gateway of progress. So when this happens, we need to have the poise and grace to accept that we, who might not be part of that, are not 'left behind' but were part of the tree necessary to produce the fruit. In the end, our happiness is in our own being and not in the function or progress of another.

Perhaps you have grasped a sense of wonder about life, from the molecule to the galaxies, from the bacteria to Sapiens, from eternal time before to our era of time to future time eternal. If we choose, we can see that at every level of existence, God has a place. God doesn't compete with our place in the world and our wonderings about how it works. He is the link between the tangible physical universe and, I argue, the intangible forces that make it all happen (referring not to the quantum but to what made the quantum). No matter how smart we get about how things work, we need to open our being to the greater being from whence we came, of which we are a part, and to where we shall go.

Our Story, A Prayer

The creation of consciousness is a story of love — our completion, the indescribable beauty, the awareness of the universe and God, the union of our being with everything else, the joy of humanity as well as the agony all in a moment and in an eternity, the hope of our souls, the experiences we glimpse. Knowledge has been unable to explain love, not referring to the description of the parts of the brain that spark with love, nor the biochemical reactions and hormones that mediate it, and neither the psychosociological and evolutionary expediency of love's existence, but in the deeper layers of understanding. Love and life have similar mysteries about them in that we don't really know why they exist. As we touched upon in this book, life could slide into techno, essentially leaving 'life' behind but still allowing for existence. Love and its emotional comrades seem to be necessary to connect cerebral processes in order to make decisions and so, regardless of romantic notions about love, the emotional network is still necessary for us to live.

So now we combine all these ideas and we see the possible beauty, we see our species attaining godhead, and we see the evil inherent in our being that limits us and we ask, "What if we just destroyed all the bad in us?" We would have to acknowledge that the good and the evil, the love and the ugly, the regeneration and the cancer, are intrinsically intertwined. This is part of our frustration, that part of us is destructive — to ourselves and the Earth — while at the same time capable of envisioning the attainment of our fantastic imaginations. The small part of our dark side adheres to the drives of evolution — that merciless, insidious and ever-present push to compete, that insists that we play our part for a while then go extinct, making room for the next hominid. There will come a time when the last book is written, the final epitaph, the book at the other end of the Sapiens library from the very first one, *Gilgamesh*, that completes our species' story. Will we look to the heavens and say goodbye with comfort and warmth in our souls, knowing that we played our part well, or will we be filled with regret, having been slaves to forces beyond our control or understanding? Unlike Neanderthal who likely expired unceremoniously on the shores near Gibraltar, will we graciously bequeath our souls to God and our molecules back to the Earth when the time comes to pass the baton to the next species?

Possibilities

So what does the future hold? The most scientific answer, at this time, is that no one knows. If we consider that history is usually the best indicator of the future, then *Homo sapiens* is destined to follow all his cousins into extinction and a new Homo will arise. The rewriting of history that could occur is that the new Homo won't arise from Africa, where all the previous Homos came from, but from a new locus on Earth. Those societies that live without the embrace/prison of modern technology, such as tribes in New Guinea, the Hadza of Tanzania, Bedouin nomads of the Sahara desert, the Inuit (Eskimo) of the Arctic, or Australian Aborigines who live traditionally, could be the survivors to hold the gene pool of the future. We are at a unique point in history in that cataclysmic change on a global scale, this time due to Sapiens activity and not an astronomical one (asteroid impaction) or geophysical one (ice age, volcanoes), could effect another massive global extinction, which would be the fifth big one since the beginnings of life on Earth. If this happens, then the cards are thrown into the air and any number of possibilities exists. If events start to unfold that precipitate huge changes on Earth, like the start of an ice age climate, then the ecosystem necessary for comfort and food that sustains Sapiens might no longer exist. No habitat, no Sapiens. What next?

The classic succession story is the one described in *Planet of the Apes*, whereby the next most complex primate experiences neurological growth to slide into technological first place. If primates don't do this, then another mammalian would be in line to develop opposing digits and cerebral capacities to advance to technological culture.

If no mammalians survive then the top spot is up for grabs by any species that can evolve, including reptile or avian, or a whole new category of creatures. Reptiles (dinosaurs) and birds already had their time of being the dominant ones on Earth and didn't venture to the stars, so chances are that a second time around wouldn't be any more productive.

Not that it's looking imminent, but if most of advanced life is wiped out, then we would have to start all over from the beginning, and if life has a path that favours a primate successor, the evolution could take 100 million years. But by starting life all over, we open up the possibilities to an entirely new family of species to come into existence, like a walking, thinking plant, or a hybrid of reptile and avian, or some other classification that has never been before, like a jellyfish that uses telepathic energy instead of opposing digits. Perhaps a species that has arisen on distant planets, and which purportedly visits us occasionally, is waiting for that development to occur, or for the right timing to trigger it.

Another possibility, even less likely, is that the big urge for life, pushing it to arise out of the rocks, and all life striving for more/bigger/better only has so much time. If life on Earth is totally destroyed, then we might put the great sequence of the evolution of life so far back, by billions of years, to the time before microbes, so that life never gets another chance on this planet to re-attain our current status quo before the dying Sun embraces Earth.

Figure 31

The End

Finale

What we have been part of up until now has worked for us in the sense of species survival, but now we are coming to a gateway of possibilities, where attaining our maximum potential might not be via evolution's path. There are things that we can do. One person on a good path times millions results in millions on a good path. Our creative technological brilliance will be stimulated, as in times of necessity and war, to open up possibilities to attempt to answer big questions of existence — even survival. If we believe that a balance is necessary between the intuitive and spiritual with the tangible and logical, then intelligence will need to walk with compassion.

Soft Touch

Working as a veterinarian, I would sometimes attend to a newborn foal. It is crucial for a foal to drink milk as soon as possible, but it takes quiet gentleness to attain that goal. Without some assistance, an exhausted foal could lie for long periods, and with too much interference, the foal could get irritated and struggle, using up essential energy.

One technique was to have someone hold the mare still and another person line the foal up with his[13] rear to the mare's head and his head in front of her back leg. This presents the dark underbelly of the mare to the foal, which is part of the trigger for his innate senses to stimulate him to nuzzle and bunt with his nose, which in turn causes a milk letdown from the mare.

The person handling the foal has to be simple and brief in their guiding actions, almost not there, more like a breeze, so that when the foal wanders, he is gently brought back to the starting position to look again at the great potential for sustenance and survival. He is presented with the opportunity and then is allowed the freedom to find his own way, but he has to find out for himself how to proceed — no one can force him, and force actually diminishes the chances of success.

Everyone believes in something, and most people believe in something spiritual. For many of the nonGod people there is the possibility that what they see as God, they don't call God. If it is personal it might be like All-Who-Is, or something less personal but no less awesome, like Truth, the essence of divinity or universal completion.

13. 'Foal' refers to both genders; a colt is a male and a filly is a female, but for conversation's sake, this foal is termed 'he'.

It is amazing to learn that the holy books and scholars of Jewish, Christian and Muslim faiths all tell us that God is beyond words, to the extent that He is even called Nothing — as Karen Armstrong explains in her book *The Case for God*. The holy men who called Him that were saying that He is 'no thing', meaning that He is not material. As our observations and language exist in the material world, words are not capable of describing the indescribable. It is like showing the foal where the milk is — the words guide us to a path, but the discovery is ours alone.

In the past 500 years or so, logic and science lifted us out of somnolence and ignorance, and in that transition at least two processes were at work: the first, where spirituality, encased by formal religion, was seen to be oppressive, misinformed and self-righteous; and the second, where logic illuminated our awakening to the functioning of the universe. In that grand process, God was either engulfed in the science and scrutinized logically, or considered outside of logic's sphere and disregarded as not provable and so unworthy of consideration. Neither path leads to God.

It would have been a fine line, the razor's edge, for us to walk on a path to God, allowing Him to be indefinable and beyond our comprehension, yet still seeking to know everything about Him and the universe. We hadn't the ability and the balance to do that in ancient times, when everything was so new. Today, we have that ability and insight, and we need to restore the balance. To this end, much of our self-advice would be in the intangible arena, taking steps to balance our existence with increased compassion, and to consider a more universal and benevolent being to sit on the throne of our communal minds:

Love one another.
Be generous.
Stop the madness of the frenetic pursuit of 'happiness'.
Know that your body is temporary and your spirit eternal.

Rediscover patience, tolerance, kindness and all the other consequences of love.
Love first, understand later.
Take care of the Earth in your yard and neighbourhood as if it was part of you and your children.

In better equilibrium, we can put our minds and hearts together to unravel the mysteries of our universe and claim our destiny with the stars, instead of clubbing each other and the biosphere to death over territory or food, or soon, very soon, over water, and after that, over oxygen.

Finally

If God *is*, and we trust the integrity and insights of good people who devote their lives to God, we can receive the clear message that "God loves you". For the God people, this is a simple truth that warms the soul. For the nonGod people, this is best taken as a gift-offering without time limit, without obligation.

God loves you.
That's the spiritual part.

For our short sojourn on Earth, we command molecules that come and go from our bodies, interacting and sharing with each other and the Earth's preciously thin crust. These molecules compose our physical presence, they include the basis of our emotional and mental well-being, and they touch upon aspects of soul. Within that realm, the forces of evolution push us every day, and yet now that we are aware of our existence at progressively new levels, our neuroplastic minds can respond in new ways.

Perhaps a new breed or species of human is coming into being.
That's the physical part.

Each realm, the spiritual and the physical, beckons us to proactively take part.

How you choose to respond is in your hands.

Bibliography

Alvord, Katie. *Divorce Your Car!* Gabriola Island, Canada: New Society Publishers, 2000.

Armstrong, Karen. *The Case for God*. Toronto, Canada: Vintage Canada, 2010.

Bourla, David. *Push*. Film. Directed by Paul McGuigan. Universal City, USA: Summit Entertainment, 2009.

Bradshaw, J. *Healing the Shame that Binds You*. Deerfield Beach, USA: HCI, 2005.

Castaneda, Carlos. *The Teachings of Don Juan: A Yaqui Way of Knowledge*. Berkeley, USA: U of California Press, 1968.

Christopher, John. *No Blade of Grass*. New York, USA: Simon & Schuster, 1956.

Clarke, Arthur C. *2001: A Space Odyssey*. New York, USA: New American Library, 1968.

Cooper, James Fenimore, Philip Dunne, John L. Balderston, Paul Perez, and Daniel Moore. *The Last of the Mohicans*. Film. Directed by George B. Seitz. Los Angeles: Reliance Pictures, 1936.

Ehrmann, Max. *Desiderata*. Birmingham, England: Brooke House, 1972.

Finlayson, Clive. *The Humans Who Went Extinct*. Oxford, England: Oxford University Press, 2009.

Hawking, Stephen. *A Brief History of Time*. New York, USA: Bantam Dell, 1996.

Heminway, John. *Stress: Portrait of a Killer*. Documentary. Directed by John Heminway. Washington, USA: National Geographic, 2008.

Keeley, Lawrence H. *War Before Civilization*. Oxford, England: Oxford University Press, 1996.

Kiko, Michio. *Physics of the Impossible*. New York, USA: Anchor Books, 2008

"Lesions and Behavior Associated with Forced Copulation of Juvenile Pacific Harbor Seals (Phoca vitulina richardsi) by Southern Sea Otters (Enhydra lutris nereis)"; Harris, HS, et al, *Aquatic Mammals*. 2010;36(4):331-341(11).

Lewis-Williams, David. *The Mind in the Cave*. London, England: Thames & Hudson Ltd, 2002.

Mitchell, Stephen. *Gilgamesh*. New York, USA: Free Press, 2004.

Potter, Christopher. *You Are Here: a Portable History of the Universe*. Toronto, Canada: Knopf Canada, 2009.

Wilson, Michael and Rod Serling, adapted from novel by Pierre Boulle. *Planet of the Apes*. Film. Directed by Franklin J. Schaffner. Los Angeles, USA: Twentieth Century Fox, 1968.

Wright, Robert. *The Evolution of God*. New York, USA: Little, Brown, and Company, 2009.

Wyndham, John. *The Midwich Cuckoos*. London, England: Michael Joseph Publishers, 1957.

Suggested Reading

Anthony, David W. *The Horse, the Wheel and Language.* Princeton, USA: Princeton University Press, 2007.

Bahn, Paul G. ed. *Written in Bones: how human remains unlock the secrets of the dead.* Toronto, Canada: Firefly Books Ltd, 2002.

Masello, Robert. *Robert's Rules of Writing.* Cincinnati, USA: Writer's Digest Books, 2005.

Pollard, Justin. *The Story of Archeology in 50 Great Discoveries.* London, England: Quercus, 2007.

This is the book containing knowledge of the Tuatha de Danaan, that started my book, and it was found in a corner of a little shop in the seaside town of Dingle, Ireland:
May, C.P.R. *The Discovery of the Dingle Diamond.* Isle of Arran, Scotland: The Banton Press, 2006.

That book was found in the quest of visiting the Great Blasket Island nearby, to witness the birthplace of the author of this book, easily one of the most profound books I've read:
O'Crohan, Tomas. *The Islandman.* Oxford, England: Oxford University Press, 1951.

So, in the end, one book has begat another.

Index

About the Author

Brett Hayward has spent his lifetime working with animals as a veterinarian and otherwise, learning and wondering about life, death and the existential questions in-between. An activist on social issues, particularly homelessness, mental health and drug addiction, he has worked with several agencies, including abroad with a missionary organization, and now lives on Vancouver Island.